Springer Proceedings in Complexity

Springer Proceedings in Complexity publishes proceedings from scholarly meetings on all topics relating to the interdisciplinary studies of complex systems science. Springer welcomes book ideas from authors. The series is indexed in Scopus. Proposals must include the following: – name, place and date of the scientific meeting – a link to the committees (local organization, international advisors etc.) – scientific description of the meeting – list of invited/plenary speakers – an estimate of the planned proceedings book parameters (number of pages/articles, requested number of bulk copies, submission deadline). submit your proposals to: christopher.coughlin@springer.com

More information about this series at http://www.springer.com/series/11637

Fakhteh Ghanbarnejad • Rishiraj Saha Roy
Fariba Karimi • Jean-Charles Delvenne
Bivas Mitra
Editors

Dynamics On and Of Complex Networks III

Machine Learning and Statistical Physics Approaches

 Springer

Editors
Fakhteh Ghanbarnejad
Institute of Theoretical Physics
Technical University of Berlin
Berlin, Germany

Rishiraj Saha Roy
Max Planck Institute for Informatics
Saarbrücken, Germany

Fariba Karimi
Department of Computational Social
Science
GESIS
Leibniz Institute for the Social Science
Köln, Germany

Jean-Charles Delvenne
Université Catholique de Louvain
Louvain-la-Neuve, Belgium

Bivas Mitra
Department of Computer Science and
Engineering
Indian Institute of Technology Kharagpur
Kharagpur, India

ISSN 2213-8684   ISSN 2213-8692   (electronic)
Springer Proceedings in Complexity
ISBN 978-3-030-14685-6   ISBN 978-3-030-14683-2   (eBook)
https://doi.org/10.1007/978-3-030-14683-2

This Springer imprint is published by the registered company Springer Nature Switzerland AG.
The registered company address is: Gewerbestrasse 11, 6330 Cham, Switzerland

Preface

Recently "network science" has been bridging various disciplines like mathematics, physics, biology, chemistry, computer science, ecology, and the social sciences. This is mainly due to its wide perspective in modeling the structure and dynamics of complex systems, both natural and man-made, with different, large, or even multiple scales. Some examples include genetic networks, food web, trade networks, the World Wide Web (WWW), collaboration networks, power grids, and air traffic networks.

The primary aim of the workshop series on *Dynamics on and of Complex Networks* (DOOCN) is to systematically explore the statistical dynamics "on" and "of" complex networks that prevail across a large number of scientific disciplines. *Dynamics on networks* refers to dynamical processes which evolve on networks and their evolution, which is impacted by their underlying topology. On the other hand, *dynamics of networks* refers to the changes occurring in the topology due to various interactions. The first DOOCN workshop (DOOCN-I) took place in Dresden, Germany, 2007, as a satellite workshop of *the European Conference on Complex Systems* (ECCS). After the success of DOOCN-I, new editions were organized as satellites of *ECCS/CCS* in Jerusalem (2008), Warwick (2009), Lisbon (2010), Vienna (2011), Barcelona (2013), Lucca (2014), Amsterdam (2016), and Thessaloniki (2018) and also as satellites of *the International School and Conference on Network Science* (NetSci) in Zaragoza (2015) and Indianapolis (2017). Details of the workshop series are available at doocn.org.

Eminent speakers of the workshops in recent years included (alphabetically ordered) Katharina Zweig (TU Kaiserslautern, Germany), Renaud Lambiotte (University of Oxford, UK), Yamir Moreno (University of Zaragoza, Spain), Sayan Pathak (Microsoft Research Redmond, USA), Ginestra Bianconi (Queen Mary University of London, UK), Constantine Dovrolis (Georgia Institute of Technology, USA), Martin Rosvall (Umea University, Sweden), Frank Schweitzer (ETH Zurich, Switzerland), Krishna Gummadi (Max Planck Institute for Software Systems, Germany), Ciro Cattuto (ISI Foundation, Italy), Markus Strohmaier (RWTH Aachen University, Germany), and Matthieu Latapy (UPMC, Paris).

Notably, the workshop organizing committee has published two book volumes from the selected talks of the series (2009: https://goo.gl/tmqPYm) and (2013: https://goo.gl/GQkfEp). The first volume aimed to show how complex network theory is being successfully used by researchers to tackle numerous difficult problems in various domains and included three parts addressing applications of complex networks in biological, social, and information sciences. The second volume aimed to put forward burgeoning multidisciplinary research contributions that combine methods from computer science, statistical physics, econometrics, and social network theory toward modeling time-varying social, biological, and information systems. This volume included three parts: (1) online social media, the Internet, and the WWW, (2) community analysis, and (3) diffusion, spreading, mobility, and transport.

The third book volume is the present one, edited by the DOOCN 2017 organizers, and aims to focus on this specific topic: "Machine Learning and Statistical Physics". Recently, machine learning (ML) techniques have been used to model dynamics of massive complex networks generated from big data and various functionalities resulting from the networks. It has become clear from the past DOOCN workshop editions that modeling large-scale dynamic networks, such as mobile adhoc networks and societal opinion networks, has gained enormous relevance in the landscape today. The advent of big data technologies, which allow effective acquisition and processing of massive amounts of unstructured data, further promises to improve the effectiveness and cross-fertilization of ML and network science. This motivated us to focus on this area of significant recent interest in our last workshop editions. A key feature of the DOOCN workshops is that each year, an exciting theme is chosen; some recent themes include "Big Data" (2014), "Computational Aspects of Big Data" (2015), "Mining and learning for complex networks" (2016), "Machine learning and statistical physics" (2017), and "Machine learning for complex networks" (2018). This volume presents a mix of very relevant reviews of important works in the field and gives the reader an up-to-date picture of the state of the art. This edition also contains independent research reports.

This book volume consists of three major parts. The contributions in the first part focus on network structure, with three chapters. In the first chapter "An Empirical Study of the Effect of Noise Models on Centrality Metrics", Sarkar et al. conducted an empirical study of how different noise models affect the network structure, precisely the ranking of centrality metrics. The analysis presented in this chapter reveals that the stability of the ranking varies according to the structure of the network, the noise model used, and the centrality metric to be computed. In the second chapter "Emergence and Evolution of Hierarchical Structure in Complex Systems", Siyari et al. investigated the following key questions in the context of modeling the emergence and evolution of hierarchical structure in complex systems: (a) How do key properties of emergent hierarchies, like depth of the network, centrality of each module, and complexity of intermediate modules, depend on the evolutionary process that generates the new targets of the system? (b) Under what conditions do the emergent hierarchies exhibit the so-called hourglass effect? (c) Do intermediate modules persist during the evolution of hierarchies? In the third chapter

"Evaluation of Cascading Infrastructure Failures and Optimal Recovery from a Network Science Perspective", Warner et al. reviewed the network science literature in order to create a hypothesis for the recovery of infrastructure systems. They represented the cascade of infrastructure systems through networks and simulated perturbations within singular and discussed how these impact multiple layers of an interconnected network.

Part II of the book volume focuses on network dynamics and it spans over four chapters. In the fourth chapter "Automatic Discovery of Families of Network Generative Processes", Menezes and Roth first reviewed the principles, efforts, and emerging literature in this direction, which is aligned with the idea of creating artificial scientists. Next, the authors developed an approach to demonstrate the existence of families of networks that may be described by similar generative processes. In the fifth chapter "Modeling User Dynamics in Collaboration Websites", Kasper et al. presented several approaches to deepen the understanding of user dynamics in collaborative websites. Inevitably, these approaches are quite heterogeneous and range from simple time-series analysis toward the application of dynamical systems and generative probabilistic methods. In the sixth chapter "Interaction Prediction Problems in Link Streams", Arnoux et al. addressed the problem of predicting future interactions, which is traditionally addressed by merging interactions into a graph or a series of graphs, called snapshots. However, in this chapter, authors formalized interactions within the link stream framework, which makes it possible to fully capture both temporal and structural properties of the data. In the seventh chapter "The Network Source Location Problem in the Context of Foodborne Disease Outbreaks", Horn and Friedrich introduced the source identification problem in the context of foodborne disease outbreaks based on basic practical knowledge of food supply networks and the foodborne disease contamination process.

The third part of the book volume focuses on theoretical models and applications with three chapters. In the eighth chapter "Network Representation Learning using Local Sharing and Distributed Graph Factorization (LSDGF)", Pandey proposed a distributed algorithm for network representation learning (NRL), which learns matrix factorization of a given network in which a node utilizes only the information available at its neighboring nodes and connected nodes for exchanging feature vectors dynamically. The performance of the proposed algorithm is evaluated by the learning of first-order proximity, spectral distance, and link prediction. In the ninth chapter "The Anatomy of Reddit: An Overview of Academic Research", Medvedev et al. explored one of the most popular social media platforms, Reddit. They developed a suite of methodologies to extract information from the structure and dynamics of the Reddit system. In the tenth chapter, "Learning Information Dynamics in Online Social Media: A Temporal Point Process Perspective", Samanta et al. proposed two models: (1) a probabilistic linear framework that unifies influence of different factors contributing to the popularity of an item and inter-item competitions and (2) a more generic model, with a deep probabilistic machinery that unifies the nonlinear generative dynamics of a collection of diffusion processes, and inter-process competition.

This cross-disciplinary collection of articles highlights the bridging of a variety of scientific branches. The chapters are designed to serve as the state of the art not only for students and new entrants but also for experts who intend to pursue research in this field. All the chapters have been carefully peer-reviewed in terms of their scientific content as well as readability and self-consistency. We would like to thank the authors for their contributions and their careful consideration of the editorial comments. Moreover, we acknowledge all the reviewers as listed below for their constructive criticisms, comments, and suggestions, which have significantly improved the quality of the chapters. Also, we acknowledge Satadal Sengupta for maintaining the electronic platform for the manuscript submission, management and monitoring. Finally, we are extremely grateful to the entire support team from Springer for their help that made the timely publication of this volume possible.

Berlin, Germany   Fakhteh Ghanbarnejad
Saarbrücken, Germany   Rishiraj Saha Roy
Köln, Germany   Fariba Karimi
Louvain-la-Neuve, Belgium   Jean-Charles Delvenne
Kharagpur, India   Bivas Mitra

List of Reviewers (Alphabetically Ordered by Last Names)

Edmund Barter, Vitaly Belik, Parantapa Bhattacharya, Sanjukta Bhowmick, Tanmoy Chakraborty, Jean-Charles Delvenne, Mauro Faccin, Saptarshi Ghosh, Martin Gueuning, Jean-Loup Guillaume, Flavio Iannelli, Lucas Jeub, Suman Kalyan Maity, Alexey Medvedev, Andrew Mellor, Subrata Nandi, Camille Roth, Koustav Rudra, Michael Schaub, Sandipan Sikdar, and Simon Walk.

Contents

Part III   Theoretical Models and Applications

Part I
Network Structure

An Empirical Study of the Effect of Noise Models on Centrality Metrics

Soumya Sarkar, Abhishek Karn, Animesh Mukherjee, and Sanjukta Bhowmick

Abstract An important yet little studied problem in network analysis is the effect of the presence of errors in creating the networks. Errors can occur both due to the limitations of data collection techniques and the implicit bias during modeling the network. In both cases, they lead to changes in the network in the form of additional or missing edges, collectively termed as noise. Given that network analysis is used in many critical applications from criminal identification to targeted drug discovery, it is important to evaluate by how much the noise affects the analysis results. In this paper, we present an empirical study of how different types of noise affect real-world networks. Specifically, we apply four different noise models to a suite of nine networks, with different levels of perturbations to test how the ranking of the top-k centrality vertices changes. Our results show that deletion of edges has less effect on centrality than the addition of edges. Nevertheless, the stability of the ranking depends on all three parameters: the structure of the network, the type of noise model used, and the centrality metric to be computed. To the best of our knowledge, this is one of the first extensive studies to conduct both longitudinal (across different networks) and horizontal (across different noise models and centrality metrics) experiments to understand the effect of noise in network analysis.

Keywords Noise models in networks · Centrality metrics · Accuracy of analysis

S. Sarkar · A. Mukherjee
IIT Kharagpur, Kharagpur, West Bengal, India
e-mail: soumya015@cse.iitkgp.ernet.in; animeshm@cse.iitkgp.ernet.in

A. Karn
NIT Durgapur, Durgapur, West Bengal, India

S. Bhowmick (✉)
University of North Texas, Denton, TX, USA
e-mail: sanjukta.bhowmick@unt.edu

© Springer Nature Switzerland AG 2019   3
F. Ghanbarnejad et al. (eds.), *Dynamics On and Of Complex Networks III*,
Springer Proceedings in Complexity, https://doi.org/10.1007/978-3-030-14683-2_1

1   Introduction

In recent years, network analysis has become an important mathematical tool for studying the interactions of entities in complex systems. The entities are represented as the vertices and their dyadic relations are represented as edges. The structural properties of the network provide insights to the characteristics of the underlying system. For example, high centrality vertices point to important proteins in protein–protein interaction networks [5] and groups of tightly connected vertices, or communities represent groups of friends in social networks [19].

An important yet little studied problem in network analysis is the effect of the presence of errors in creating the networks. The errors primarily occur at two stages: First, they occur when collecting real-world data. The measurements of any physical system inherently include some degree of error, which is propagated to the network model. Second, the errors occur during the creation of the network model. The inter-relations (here edges) are often determined based on the subjective evaluation of the modeler. For example, in a gene correlation network, two vertices (genes) are connected by an edge only if the correlation between the two genes is higher than a specified threshold. However, due to the absence of a standard value, this threshold is decided by the person creating the model.

These errors in data collection and modeling are manifested as structural changes in the network, in the form of additional or missing edges, collectively termed as *noise*. An important question is by how much this noise affects the analysis results. In particular, since network analysis is used in many critical applications from criminal identification [12] to targeted drug discovery [3], a drastic change in the accuracy can lead to serious consequences.

Overview In this paper we present an extensive empirical study of how different types of noise affect real-world networks. Specifically, we apply four different noise models. These are random deletion of edges (edge deletion), random addition of edges (edge addition), swapping the end-points of a pair of edges (edge swap), and overlaying the network with a random graph such that only the edges present in either the original or the random network are kept (edge XOR). Each of these noise models is applied to a suite of nine networks, with different levels of perturbations to test how the ranking of the top-k centrality vertices changes.

We measure the change in the ranking using the Jaccard index (JI). We test how the following centrality metrics, *degree centrality*, *betweenness centrality*, and *closeness centrality*, are affected by these noise models. While there have been several studies [4, 6, 18] conducted on individual noise models and how they affect network properties, to the best of our knowledge, this is one of the first studies to conduct both a longitudinal (across several different networks) and a horizontal (across noise models and centrality metrics) evaluation of the effect of noise in networks. Some of the results that we observe through these experiments are:

- *Variations in Noise Models:* Edge swap produces the highest average JI, i.e., the least amount of change in the vertex ranking. This is followed by, in order of highest to lowest average JI, edge deletion, edge XOR, and finally edge addition.
- *Variations in Network Structure:* Networks from the technological domain, particularly the autonomous network AS2 and the peer-to-peer network P2P, show the most stability, i.e., high JI across all noise models. The network of the power grid of the Western United States shows the least stability with lowest JI across all noise models. The other networks show high stability for edge swap and edges deletion and low stability for the other two models.
- *Variations in Centrality Metrics:* Of the three centrality metrics, degree centrality was the most stable, and betweenness centrality was the least stable.

The remainder of this paper is organized as follows: In Sect. 2, we present our experimental methodology along with the datasets and definitions of the centrality measures. In Sect. 3, we describe the effect of the noise models and provide a summary of our observations. In Sect. 4, we discuss related research in this domain. We conclude in Sect. 5 with an overview of our future research plans.

2 Experimental Setup

In this section we provide a brief description of our experimental setup, including description of the networks used, the definitions of the centrality metrics, and an overview of how we conducted the experiments.

2.1 Test Suite of Networks

We consider the following nine networks that were collected from public repositories [8, 13]. We group these networks into three categories: (1) technological networks that are formed through interconnections of routers or peer-to-peer networks, (2) social networks that are formed through collaborations, chats, or linking between blogs, and (3) miscellaneous networks formed from other varied applications, including biological networks, software networks, and power grids. A summary of the networks, their descriptions, and sizes is given in Table 1.

2.2 Centrality Metrics

In our experiments we study the stability of the following centrality metrics. Given a graph $G(V, E)$, with $|V|$ vertices and $|E|$ edges, the metrics are computed as:

Table 1 Description of the nine networks in the test suite

Network	Description	Node	Edges
Technological networks			
AS	Network of routers obtained from University of Oregon Route Views Project	6474	13,895
CA	Network of routers obtained from Center for Internet Data Analysis (CAIDA)	16,493	66,744
P2P	Gnutella peer-to-peer file sharing network	26,518	65,369
Social networks			
APH	Collaboration network of authors of papers posted in arXiv's Astrophysics	16,046	121,251
AnyBeat	Social network where users connect anonymously	12,645	67,053
Blog	Network of front-page hyperlinks between blogs related to the 2004 US election	1224	19,025
Miscellaneous networks			
PW	Network of power grid of the Western United States	4941	6594
BIO	Protein–protein interaction network	7393	25,569
SW	Dependency network of classes in JUNG and javax	6120	50,535

From left to right, the columns are: abbreviation of the network name, short description of the network, the number of vertices, the number of edges, and the global clustering co-efficient

Degree centrality, $D(v)$, of a vertex v measures the number of its neighbors. *Closeness centrality* of a vertex v measures its average distance from all other vertices in the network. It is computed as $CC(v) = \dfrac{|V|}{\sum\limits_{s \neq v \in V} dis(v,s)}$, where $dis(v, s)$ is the length of a shortest path between v and s.

Betweenness centrality of a vertex v is defined as the fraction of the total number of shortest paths that pass through the vertex. It is computed as $BC(v) = \sum\limits_{s \neq v \neq t \in V} \dfrac{\sigma_{st}(v)}{\sigma_{st}}$, where σ_{st} is the total number of shortest paths between s and t, and $\sigma_{st}(v)$ is the total number of shortest paths between s and t that pass through v.

2.3   Methodology

To test the stability of the networks and centrality metrics under different noise models, we perform the following experiment. We first compute the centrality values of each network, and rank the vertices from highest to lowest centrality values. There can be potentially different ranking for each centrality metric. We test the effect on rank, rather than the value of the centrality, because most applications, such as information spreading or vaccination, require finding the high ranked vertices and do not require their exact value. We apply the noise models to the networks over different levels of perturbations, and compute the centrality, and subsequent ranking on the perturbed network.

We then use Jaccard index (JI) to measure how many of the top-k vertices from the original ranking are retained. The JI of two sets A and B is given by $\frac{A \cap B}{A \cup B}$. The highest value is 1, when both sets A and B are the same and the lowest value is 0, when the sets A and B do not have any common elements. We test JI for the top ranked 5, 10, 25, and 50 vertices. Since the noise models are stochastic, test is repeated 5 times, and the average JI over the five tests is reported for each tuple of noise model, network types, centrality metric, and perturbation level.

3 Empirical Results

We describe the stability results as per our experiments on the four noise models. We test one model for only addition of edges, one for only deletion, and two models that involve both addition and deletion of edges. Addition and deletion are the units of change in a network, so we study them individually. Variations of the other two models have been used in [2] and [6]. We present the detailed results for each model separately in Sects. 3.1–3.4 and summarize our findings from all these experiments in Table 2.

3.1 Edge Addition

In this noise model we add edges to the network. We select a pair of vertices from the set V with probability $\frac{p}{|V|}$. If the edge is not already part of the network we add it to the network. Figures 1, 2, and 3 show how the top-k centralities change as the value of p is increased. The values of p ranged from 0.5, 1.5, 2.5, 3.5, and 4.5. As

Table 2 Average across centrality metrics for each noise model for each network

Network	Edge deletion	Edge addition	Edge swap	Edge XOR
Technological networks				
AS2	0.8	0.59	0.9	1
CA	0.7	0	0.96	0
P2P	0.68	0.61	1	0.80
Social networks				
AnyBeat	0.8	0	1	0
APH	0.6	0	1	0
Blog	0.97	0.09	0.93	0.03
Miscellaneous networks				
BIO	0.78	0	0.88	0
PW	0.47	0.02	0.34	0.11
SW	0.95	0.13	0.9	0.21

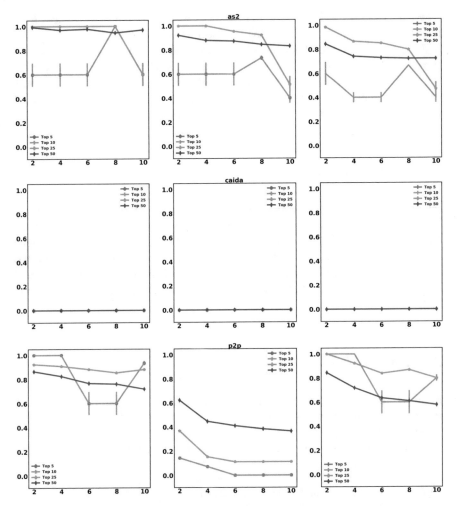

Fig. 1 Effect of edge addition on technological networks. X-axis: perturbations values; Y-axis: Jaccard index. Left: degree; middle: betweenness; right: closeness. Top to bottom: AS, CAIDA, and P2P

can be seen, apart from AS2 and P2P, all the other classes of networks exhibit very low JI for every perturbation and every value of k. Thus *edge addition even at small levels of perturbation can significantly change the ranking.*

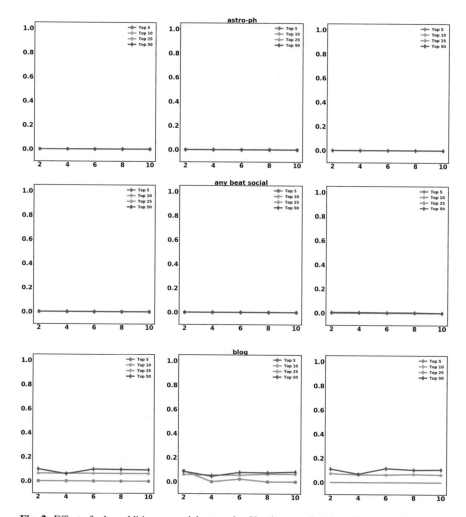

Fig. 2 Effect of edge addition on social networks. X-axis: perturbations values; Y-axis: Jaccard index. Left: degree; middle: betweenness; right: closeness. Top to bottom: APH, AnyBeat, and Blog

3.2 Edge Deletion

In this noise model we delete edges from the network. We select an existing edge from the set E and remove it with a probability of p. In our experiments we set p to range from 2, 4, 6, 8, and 10% of the edges. Figures 4, 5, and 6 show how the top-k centralities change as the value of p is increased.

We observe that in contrast to the edge addition model, the JI values for degree and closeness centralities are generally high for all networks. The behavior of the JI values of betweenness centrality varies from being 1, i.e., ranking unaffected (Blog)

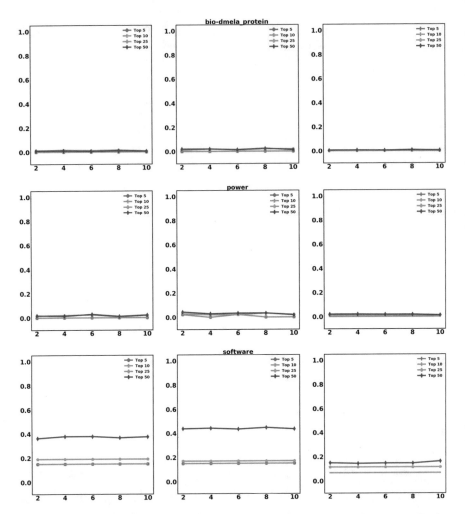

Fig. 3 Effect of edge addition on miscellaneous networks. X-axis: perturbations values; Y-axis: Jaccard index. Left: degree; middle: betweenness; right: closeness. Top to bottom: BIO, PW, and SW

to gradually decreasing (AnyBeat) to being 0, i.e., ranking completely changed (APH). We conclude that *uniform deletion does not significantly affect the ranking of degree and closeness centralities.*

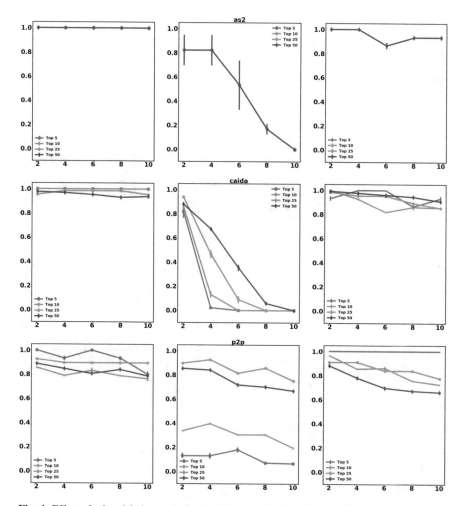

Fig. 4 Effect of edge deletion on technological networks. X-axis: perturbations values; Y-axis: Jaccard index. Left: degree; middle: betweenness; right: closeness. Top to bottom: AS, CAIDA, and P2P

3.3 Edge Swap

We now consider noise models that include both addition and deletion of edges. The first example we consider is swapping edges between two pairs of connected vertices. Let us consider two edges, (a, b) and (c, d), where none of the vertices a and b are connected to vertices c and d. In the swapping model, we disconnect the edge between a and b, and between c and d. Then, to maintain the degree of the vertices, we reconnect a with c and b with d. A version of this noise model was used in [6] to measure robustness of communities.

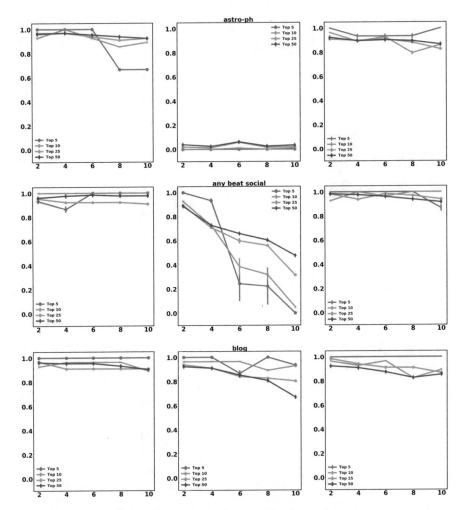

Fig. 5 Effect of edge deletion on social networks. X-axis: perturbations values; Y-axis: Jaccard index. Left: degree; middle: betweenness; right: closeness. Top to bottom: APH, AnyBeat, and Blog

Figures 7, 8, and 9 show how the top-k ranking changes as the value of p, the percentage of edge selected, is increased. Values of p are 2, 4, 6, 8,and 1. Due to the characteristics of the model, the ranking of degree will remain mostly unchanged. We see that *for most networks and centrality metrics, the JI values are high indicating that swapping does not significantly perturb the centrality ranking.* The exceptions are the power network, and some cases of betweenness centrality.

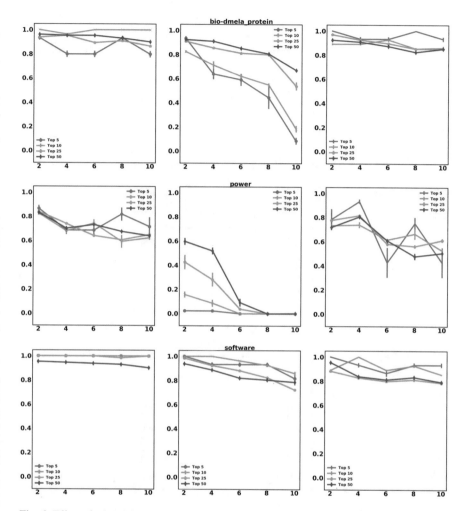

Fig. 6 Effect of edge deletion on miscellaneous networks. X-axis: perturbations values; Y-axis: Jaccard index. Left: degree; middle: betweenness; right: closeness. Top to bottom: BIO, PW, and SW

3.4 Edge XOR

In this model, we also consider both addition and deletion of edges. Here we create a random graph $R(n, p)$ with the same number of vertices, $|V| = n$ as the original graph G. We perform a perturbation in the form that if an edge (a, b) is present in both R and G, the edge is deleted from the G. However, if the edge is present in R but not in G, we add the edge to G. We term this as the XOR perturbation, because of its similarity to the boolean XOR operation (output is true if exactly one, but not both conditions are true). A version of this model was used in [1], for testing

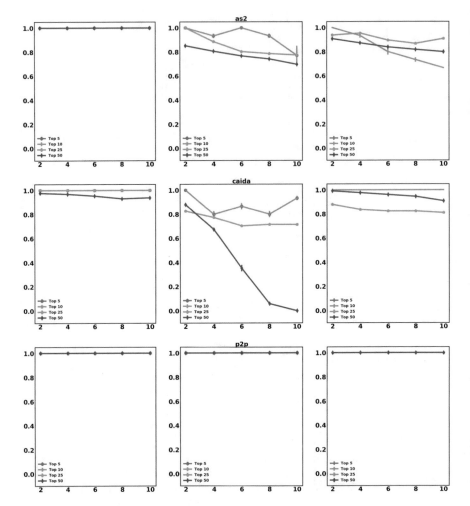

Fig. 7 Effect of edge swapping on technological networks. X-axis: perturbations values; Y-axis: Jaccard index. Left: degree; middle: betweenness; right: closeness. Top to bottom: AS, CAIDA, and P2P

robustness of the k-core. Because the networks are sparse, more edges will be added than deleted in the perturbed network. The probability with which the edges in the random network were connected is varied from 0.5, 1.5, 2.5, 3.5, and 4.5.

The results in Figs. 10, 11, and 12 show that for all networks, except for AS2 and P2P, the JI value is close to zero for all networks and centrality measures. This indicates that the *XOR model can easily disrupt the ranking of the high centrality vertices*, even for low perturbations.

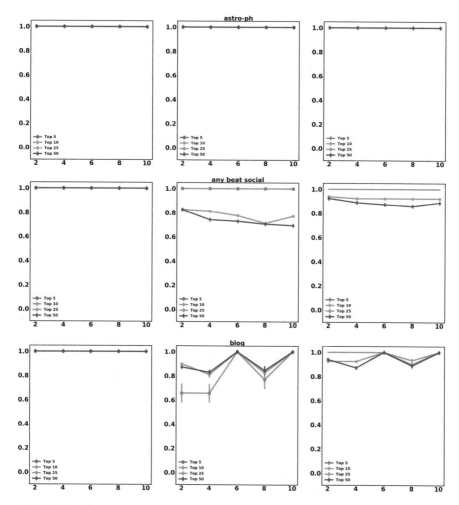

Fig. 8 Effect of edge swapping on social networks. X-axis: perturbations values; Y-axis: Jaccard index. Left: degree; middle: betweenness; right: closeness. Top to bottom: APH, AnyBeat, and Blog

3.5 Summary of the Results

In this set of experiments, we studied how different noise models affect the ranking of high centrality vertices. Table 2 summarizes the average JI for each network over all the centrality metrics, and for each noise model.

From the results it can be clearly seen that the edge deletion and edge swap affect the ranking of the high centrality vertices far less than the edge addition and edge XOR. Out of these two, the edge addition model is more disruptive. Note that almost any new edges has the potential to change the route of the shortest paths, leading

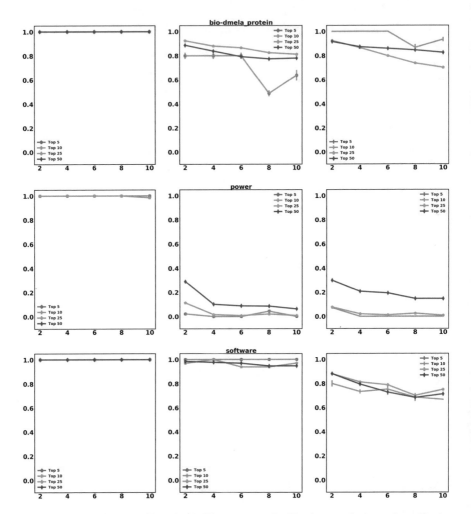

Fig. 9 Effect of edge swapping on miscellaneous networks. X-axis: perturbations values; Y-axis: Jaccard index. Left: degree; middle: betweenness; right: closeness. Top to bottom: BIO, PW, and SW

to significant changes in the betweenness and closeness centrality of the vertices. However, deletion will only affect the shortest path routes if a key edge that is part of the shortest path is deleted. We believe that this is the reason why edge addition has more impact on the centralities of the vertices.

Due to this reason, we have used lower intensity of perturbations on models that involve more edge addition (edge addition and edge XOR) than the models where edge deletion is more prominent (edge deletion and edge swap). Note that even with the higher values of perturbations, edge deletion and edge swap exhibit greater stability.

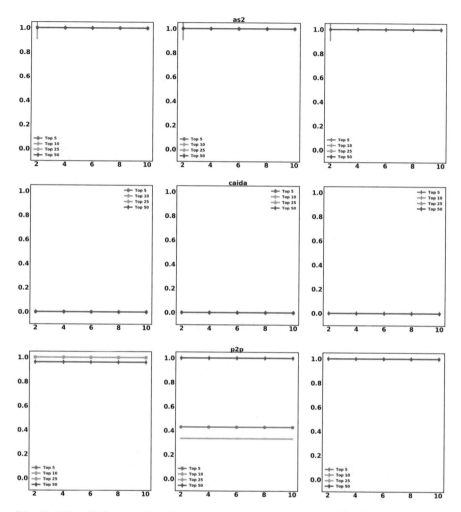

Fig. 10 Effect of XOR edge perturbation on technological networks. X-axis: perturbations values; Y-axis: Jaccard index. Left: degree; middle: betweenness; right: closeness. Top to bottom: AS, CAIDA, and P2P

We also observe that the effect of noise also depends on the network. For example, technological networks such as AS2 and P2P exhibit high JI values and are even less affected by the edge addition and edge XOR models. On the other hand, the power network (PW) is disrupted to produce low JI values even by the relatively safe edge deletion and edge swap models. We have observed in a separate angle of research [15] that the stability of networks depends on their core periphery structure. Specifically, a tightly connected inner core can make the network more stable under perturbations.

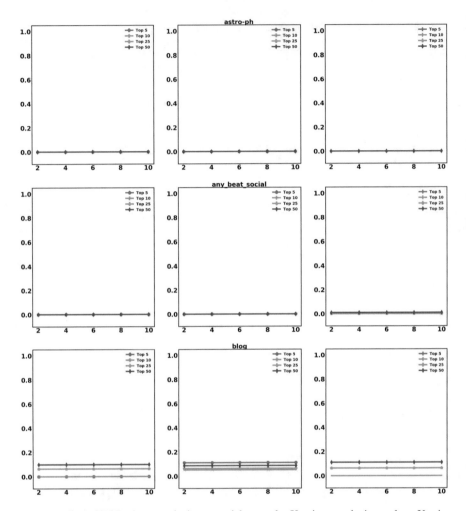

Fig. 11 Effect of XOR edge perturbation on social networks. X-axis: perturbations values; Y-axis: Jaccard index. Left: degree; middle: betweenness; right: closeness. Top to bottom: APH, AnyBeat, and P2P

Finally from the figures we see that betweenness centrality is the least stable of the three centrality metrics. We hypothesize that this is because the value of betweenness centrality not only depends on the paths passing through a vertex, but also on other shortest paths. Thus any change in network, even on paths that do not involve a vertex, can affect the centrality of the vertex.

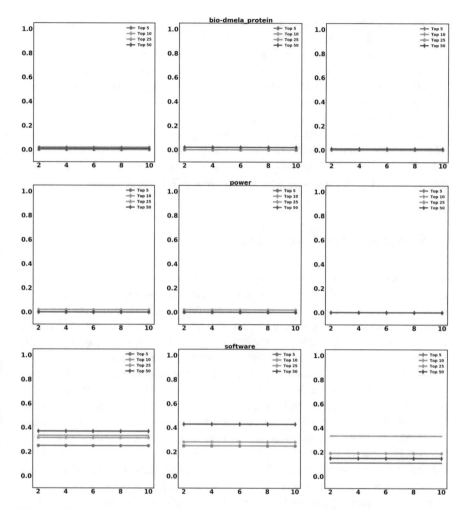

Fig. 12 Effect of XOR edge perturbation on miscellaneous networks. X-axis: perturbations values; Y-axis: Jaccard index. Left: degree; middle: betweenness; right: closeness. Top to bottom: BIO, PW, and SW

4 Related Research

The issue of noise in networks is gradually gaining prominence. In the recent years there have been several studies [4, 9, 16, 18] that explored the effect of noise on different centrality metrics. Borgatti et al. [4] used noise models of node addition/deletion and edge addition/deletion to test the change in centralities on Erdos–Renyi random networks. Wang et al. extended the research to also include false aggregation/disaggregation. False aggregation happens when two or more nodes are erroneously classified as one node. False disaggregation happens when one node is erroneously classified as separate nodes. They analyzed the effect of different centrality metrics on two real-world networks and a random network.

Segarra et al. [16] modeled noise as random fluctuations on edge weights on the graph, and theoretically measured for robustness of the centrality measures based on the noise. They showed that for the given definition of stability, the measures degree, closeness, and eigenvector are stable, while betweenness is not. This observation matches with our experimental results as well.

Other work on the effect of noise includes looking at incomplete networks with missing edges. These include solving the entity resolution problem in incomplete networks[11, 17]. Researchers [7, 10] have also looked at methods to find missing links. In [14], the authors study how missing links affect the centrality metrics.

5   Conclusion and Future Work

In this paper, we conducted an empirical study of how different noise models affect the ranking of the centrality metrics in network analysis. To the best of our knowledge this is one of the more extensive studies that look into networks from several applications, and compare against different noise models. Our results show that the stability of the ranking varies depending on the structure of the network, the type of noise model used, and the centrality metric to be computed.

Our experiments open up new questions on understanding each of these factors, which we aim to study in the future. For example, we should investigate what is the property of the network that leads to it being more resilient under perturbation. We can also explore the validity of noise models as they occur in real world, and experiment on other centrality metrics, such as PageRank. In particular, combining our observations with method of network inference for creating networks from raw data can lead into more robust and accurate network models. Finally, we also note that often centrality metrics need not be exactly computed, or even ranked. We aim to explore developing heuristics of centrality metrics that are more robust under noise.

Acknowledgements SB was supported by the NSF CCF Award #1533881 and #1725566.

References

1. Adiga, A., Vullikanti, A.K.: How robust is the core of a network? In: Proceedings of the European Conference on Machine Learning and Knowledge Discovery in Databases - Volume 8188, ECML PKDD 2013, pp. 541–556. Springer, New York (2013). https://doi.org/10.1007/978-3-642-40988-2_35
2. Adiga, A., Vullikanti, A.K.S.: How robust is the core of a network? In: Blockeel, H., Kersting, K., Nijssen, S., Železný, F. (eds.) Machine Learning and Knowledge Discovery in Databases, pp. 541–556. Springer, Berlin/Heidelberg (2013)

3. Berlusconi, G., Calderoni, F., Parolini, N., Verani, M., Piccardi, C.: Link prediction in criminal networks: a tool for criminal intelligence analysis. PLoS One **11**(4), 1–21 (2016). https://doi.org/10.1371/journal.pone.0154244
4. Borgatti, S.P., Carley, K.M., Krackhardt, D.: On the robustness of centrality measures under conditions of imperfect data. Soc Netw **28**(2), 124–136 (2006)
5. Jeong, H., Mason, S.P., Barabasi, A.L., Oltvai, Z.N.: Lethality and centrality in protein networks. Nature **411**(6833), 41–42 (2001). https://doi.org/10.1038/35075138
6. Karrer, B., Levina, E., Newman, M.E.J.: Robustness of community structure in networks. Phys. Rev. E **77**, 046119 (2008). https://doi.org/10.1103/PhysRevE.77.046119
7. Kim, M., Leskovec, J.: The network completion problem: inferring missing nodes and edges in networks. In: SDM, pp. 47–58. SIAM, Philadelphia (2011)
8. Leskovec, J., Sosič, R.: Snap.py: SNAP for Python, a general purpose network analysis and graph mining tool in Python (2014). http://snap.stanford.edu/snappy
9. Li, L., Xu, D., Peng, H., Kurths, J., Yang, Y.: Reconstruction of complex network based on the noise via QR decomposition and compressed sensing. Sci. Rep. **7**, 2045–2322 (2017)
10. Liu, J., Aggarwal, C., Han, J.: On integrating network and community discovery. In: Proceedings of the Eighth ACM International Conference on Web Search and Data Mining, pp. 117–126. ACM, New York (2015)
11. Moustafa, W.E., Kimmig, A., Deshpande, A., Getoor, L.: Subgraph pattern matching over uncertain graphs with identity linkage uncertainty. In: 2014 IEEE 30th International Conference on Data Engineering (ICDE), pp. 904–915. IEEE, New York (2014)
12. Pinto, J.P., Machado, R.S.R., Xavier, J.M., Futschik, M.E.: Targeting molecular networks for drug research. Front. Genet. **5**, 160 (2014). https://doi.org/10.3389/fgene.2014.00160. https://www.frontiersin.org/article/10.3389/fgene.2014.00160
13. Rossi, R.A., Ahmed, N.K.: The network data repository with interactive graph analytics and visualization. In: Proceedings of the Twenty-Ninth AAAI Conference on Artificial Intelligence (2015). http://networkrepository.com
14. Sarkar, S., Bhowmick, S., Kumar, S., Mukherjee, A.: Sensitivity and reliability in incomplete networks: centrality metrics to community scoring functions. In: Proceedings of the 2016 IEEE/ACM International Conference on Advances in Social Networks Analysis and Mining, ASONAM '16, pp. 69–72. IEEE Press, Piscataway (2016). http://dl.acm.org/citation.cfm?id=3192424.3192437
15. Sarkar, S., Sikdar, S., Bhowmick, S., Mukherjee, A.: Using core-periphery structure to predict high centrality nodes in time-varying networks. Data Mining Knowl Discov **32**(5), 1368–1396 (2018). https://doi.org/10.1007/s10618-018-0574-x
16. Segarra, S., Ribeiro, A.: Stability and continuity of centrality measures in weighted graphs. IEEE Trans. Signal Process. **64**(3), 543–555 (2016)
17. Verroios, V., Garcia-Molina, H.: Entity resolution with crowd errors (2015)
18. Wang, L., Wang, J., Bi, Y., Wu, W., Xu, W., Lian, B.: Noise-tolerance community detection and evolution in dynamic social networks. J. Combin. Optim. **28**(3), 600–612 (2014)
19. Wasserman, S., Faust, K.: Social Network Analysis: Methods and Applications, vol. 8. Cambridge University Press, Cambridge (1994)

Emergence and Evolution of Hierarchical Structure in Complex Systems

Payam Siyari, Bistra Dilkina, and Constantine Dovrolis

Abstract It is well known that many complex systems, both in technology and nature, exhibit hierarchical modularity: smaller modules, each of them providing a certain function, are used within larger modules that perform more complex functions. What is not well understood however is how this hierarchical structure (which is fundamentally a network property) emerges, and how it evolves over time.

We propose a modeling framework, referred to as Evo-Lexis, that provides insight to some fundamental questions about evolving hierarchical systems. Evo-Lexis models the most elementary modules of the system as symbols ("sources") and the modules at the highest level of the hierarchy as sequences of those symbols ("targets"). Evo-Lexis computes the optimized adjustment of a given hierarchy when the set of targets changes over time by additions and removals (a process referred to as "incremental design").

In this paper we use computation modeling to show that:

- Low-cost and deep hierarchies emerge when the population of target sequences evolves through tinkering and mutation.
- Strong selection on the cost of new candidate targets results in reuse of more complex (longer) nodes in an optimized hierarchy.
- The bias towards reuse of complex nodes results in an "hourglass architecture" (i.e., few intermediate nodes that cover almost all source–target paths).
- With such bias, the core nodes are conserved for relatively long time periods although still being vulnerable to major transitions and punctuated equilibria.
- Finally, we analyze the differences in terms of cost and structure between incrementally designed hierarchies and the corresponding "clean-slate" hierarchies which result when the system is designed from scratch after a change.

P. Siyari · C. Dovrolis (✉)
School of Computer Science, Georgia Institute of Technology, Atlanta, GA, USA
e-mail: payamsiyari@gatech.edu; constantine@gatech.edu

B. Dilkina
Department of Computer Science, University of Southern California, Los Angeles, CA, USA
e-mail: dilkina@usc.edu

© Springer Nature Switzerland AG 2019
F. Ghanbarnejad et al. (eds.), *Dynamics On and Of Complex Networks III*,
Springer Proceedings in Complexity, https://doi.org/10.1007/978-3-030-14683-2_2

23

Keywords Complex systems · Hierarchical structure evolution · Network science · Optimization

1 Introduction

It is well known that many complex systems, both in technology and nature, exhibit modularity: independent modules, each of them providing a certain function, are combined together to perform more complex functions [7]. Additionally, modular systems are also organized in a hierarchical way: smaller modules are used within larger modules recursively [28]. Examples of such systems exist in a wide range of environments: in natural systems, it is believed that hierarchical modularity enhances evolvability (the ability of the system to adapt to new environments with minimal changes) and robustness (the ability to maintain the current status in the presence of internal or external variations) [25, 30]. In the technological world, hierarchically modular designs are preferred in terms of design and development cost, easier maintenance and agility (e.g., less effort in producing future versions of a software), and better abstraction of the system design [27].

There are many hypotheses in the literature regarding the factors that contribute to either the hierarchy or modularity properties. Local resource constraints in social networks and ecosystems [26], modularly varying goals [15, 22, 23], selection for more robust phenotypes [12, 38], and selection for lower connection costs in a network [25] are some of the mechanisms that have been previously explored and shown to lead to hierarchically modular systems. The main hypothesis that we follow in this paper is along the lines of [25], which assumes that systems in both nature and technology care to minimize the cost of their interconnections or dependencies between modules.

An additional focus of our work is the hourglass effect in hierarchical systems. Across many fields, such as in computer networking [1], deep neural networks [19], embryogenesis [13], metabolism [35], and many others [30], it has been observed that hierarchically modular systems often exhibit the architecture of an hourglass. Informally, an hourglass architecture means that the system of interest produces many outputs from many inputs through a relatively small number of highly central intermediate modules, referred to as the "waist" of the hourglass (Fig. 1). The waist of the hourglass (also referred to as "core" in [30] as well as in this paper) includes critical modules of the system that are also sometimes more conserved during the evolution of the system compared to other modules [1, 30]. Despite recent research on the hourglass effect in different types of hierarchical systems [1, 2, 17, 30], one of the questions that is still open is to identify the conditions under which the hourglass effect emerges in hierarchies that are produced when the objective is to minimize the cost of interconnections.

In this paper, we present *Evo-Lexis*, a modeling framework for the emergence and evolution of hierarchical structure in complex systems. To develop *Evo-Lexis*, we extend a previously proposed optimization framework, called *Lexis* [33], that

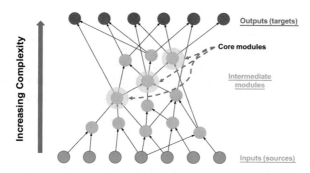

Fig. 1 A hierarchical system is represented as a directed acyclic graph in which each module is shown as a node, and the dependencies from more elementary modules to more complex modules are shown as upward edges. The hourglass effect occurs when the system of interest produces many outputs from many inputs through a relatively small number of intermediate core modules (here, highlighted nodes with transparent surroundings) [30]

was designed for structure discovery in sequential data. Lexis models the most elementary modules of the system as symbols ("sources") and the modules at the highest level of the hierarchy as sequences of those symbols ("targets"). *Evo-Lexis* is a dynamic or evolving version of Lexis, in the sense that the set of targets changes over time through additions (births) and removals (deaths) of targets. *Evo-Lexis* computes an (approximate) minimum-cost adjustment of a given hierarchy when the set of targets changes over time (a process we refer to as "incremental design"). For comparison purposes, *Evo-Lexis* also computes the (approximate) minimum-cost hierarchy that generates a given set of targets from a set of sources in a static (non-evolving) setting (referred to as "clean-slate design"). The premise behind the incremental design approach is that in practice systems are rarely designed from scratch—instead, they are incrementally modified over time to accommodate the changes (e.g., provide new outputs and potentially to support new inputs every time there is a change).

In general, a system interacts with its environment in a bidirectional manner: the environment imposes various constraints on the system and the system also affects its environment. To capture this co-evolutionary setting in *Evo-Lexis*, we study how changes in the set of targets affect the resulting hierarchy but also how the current hierarchy affects the selection of new targets (i.e., whether a new candidate target is selected or not depends on its fitness or cost—and that depends on how easily that target can be supported by the given hierarchy). By incorporating well-known evolutionary mechanisms, such as tinkering (mutation), recombination, and selection, *Evo-Lexis* can capture such co-evolutionary dynamics between the generation of new targets and the hierarchy that supports them.

The questions we focus on are:

1. How do key properties of the emergent hierarchies, e.g., depth of the network, reuse or centrality of each module, complexity (or sequence length) of intermediate modules, etc., depend on the evolutionary process that generates the new targets of the system?

2. Under what conditions do the emergent hierarchies exhibit the so-called hour-glass effect? Why are few intermediate modules reused much more than others?
3. Do intermediate modules persist during the evolution of hierarchies? Or are there "punctuated equilibria" where the highly reused modules change significantly?
4. Which are the differences in terms of cost and structure between the incrementally designed and the corresponding clean-slate designed hierarchies?

The structure of the paper is as follows: In Sect. 2, we present an overview of Lexis, the static optimization framework that serves as the main building block in Evo-Lexis.[1] In Sect. 3, we present the components of the Evo-Lexis framework, along with the metrics that we use for the analysis of evolving hierarchies. In Sect. 4, we evaluate the evolution of hierarchies under different target generation models (Fig. 2). Sections 5 and 6 present further analysis regarding the evolvability and major transitions in hierarchies produced using the most full-fledged (MRS) target generation model. Finally, Sect. 7 focuses on the comparison between clean-slate and incremental design in terms of cost and structure. In Sect. 8, we review related work in the context of Evo-Lexis. Section 9 discusses the results and presents some future research possibilities.

2 Lexis Background

In this section, we present an overview of Lexis [33], the optimization framework that we use as the main building block of the *Evo-Lexis* framework.

2.1 *Lexis-DAG*

Given an alphabet S and a set of "target" strings T over the alphabet S, we need to construct a Lexis-DAG. A Lexis-DAG D is a directed acyclic graph $D(V, E)$, where V is the set of nodes and E the set of edges, that satisfies the following three constraints[2].

First, each node $v \in V$ in a Lexis-DAG represents a string $\mathscr{S}(v)$ of characters from the alphabet S. The nodes V_S that represent characters of S are referred to as *sources*, and they have zero in-degree. The nodes V_T that represent target strings $T = \{t_1, t_2, \ldots, t_m\}$ are referred to as *targets*, and they have zero out-degree. V also includes a set of *intermediate nodes* V_M, which represent substrings that appear in the targets T. So, $V = V_S \cup V_M \cup V_T$.

---

[1]The static (i.e., non-evolving) version of the proposed modeling framework is referred to as "Lexis" and it has been published at the ACM KDD 2016 conference [33].

[2]To simplify the notation, even though D is a function of S and T, we do not denote it as such.

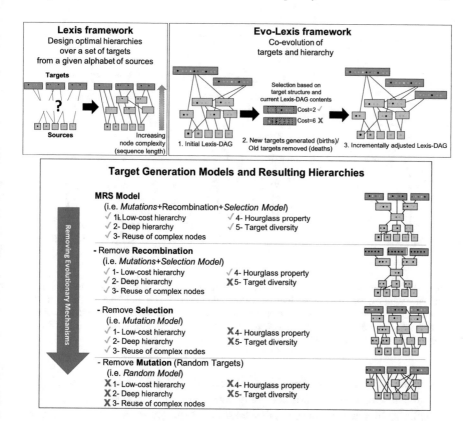

Fig. 2 Overview of this study. The *Evo-Lexis* modeling framework captures the process of incrementally designing optimized hierarchies for a time-varying set of targets. Hierarchies are modeled as *Lexis-DAGs*. We focus on key properties of the resulting hierarchies (e.g., cost, depth, and reuse of intermediate components) and on how these properties depend on the evolutionary mechanisms that generate new targets. By focusing on well-known evolutionary mechanisms such as mutations, recombination, and selection, we analyze how each of them affects the structure and evolution of the resulting hierarchies. Blue, green, and red nodes show source, intermediate, and target nodes, respectively. Colored dots represent an instance of a source node and are used to show the extent of diversity among target nodes

Second, each node in $V_M \cup V_T$ of a Lexis-DAG represents a string that is the concatenation of two or more substrings, specified by the incoming edges from other nodes to that node. Specifically, an edge $e \in E$ from node u to node v is a triplet (u, v, i) such that the string $\mathscr{S}(u)$ appears as substring of $\mathscr{S}(v)$ at index i (the first character of a string has index 1). Note that there may be more than one edges from node u to node v. The number of incoming and outgoing edges for a node v is denoted by $d_{in}(v)$ and $d_{out}(v)$, respectively.

Third, a Lexis-DAG should only include intermediate nodes that have an out-degree of at least two, $\forall v \in V_M, d_{out}(v) \geq 2$. In other words, every intermediate node $v \in V_M$ in a Lexis-DAG should be such that the string $\mathscr{S}(v)$ is reused in

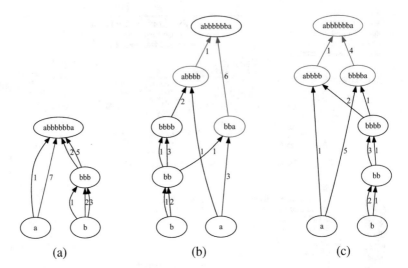

Fig. 3 Illustration of the Lexis-DAG for a single target $T = \{abbbbba\}$ and sources $S = \{a, b\}$. Edge-labels indicate the occurrence indices: **(a)** A valid Lexis-DAG having both minimum number of concatenations and edges. **(b)** An invalid Lexis-DAG: two intermediate nodes are reused only once. **(c)** An invalid Lexis-DAG: the top-layer string is not equal to the concatenation of its two in-neighbors (best viewed in color)

at least two concatenation operations. Otherwise, $\mathscr{S}(v)$ is either not used in any concatenation operation or it is used only once and so the outgoing edge from v can be replaced by rewiring the incoming edges of v straight to the single occurrence of $\mathscr{S}(v)$. In both cases node v can be removed from the Lexis-DAG, resulting in a more parsimonious hierarchical representation of the targets. Figure 3 illustrates the concepts introduced in this subsection.

2.2 The Lexis Optimization Problem

The *Lexis* optimization problem is to construct a minimum-cost Lexis-DAG for the given alphabet S and target strings T. In other words, the problem is to determine the set of intermediate nodes V_M and all required edges E so that the corresponding Lexis-DAG D is optimal in terms of a given cost function $C(D)$. This problem can be formulated as follows:

$$min_{(E,V_M)} C(D)$$
$$s.t. \ D = (V, E) \text{isaLexis} - \text{DAGfor} S \text{and} T \tag{1}$$

The selection of an appropriate cost function is somewhat application-specific. A natural cost function, as investigated in the previous work [33], is the number of

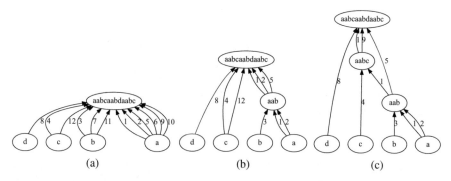

Fig. 4 Illustration of G-LEXIS algorithm given target $T = \{aabcaabdaabc\}$ and sources $S = \{a, b, c, d\}$. (**a**) Initial Lexis-DAG. (**b**) Substring aab has maximum cost reduction by reducing the number of edges in the Lexis-DAG from 12 to 9. (**c**) The substring $aabc$ has maximum cost reduction. Note how $aabc$ is partially made from the previously added substring aab. In this example, this would be the last iteration of G-LEXIS

edges in the Lexis-DAG. More general cost formulations, such as a variable edge cost or a weighted average of a node cost and an edge cost, are interesting but they are not pursued in this paper. The *edge cost* to construct a node $v \in V$ is defined as the number of incoming edges required to construct $\mathscr{S}(v)$ from its in-neighbors, which is equal to $d_{in}(v)$. The edge cost of source nodes is obviously zero. The edge cost $\mathscr{E}(D)$ of Lexis-DAG D is defined as the edge cost of all nodes, which is equal to the number of edges in D,

$$\mathscr{E}(D) = \sum_{v \in V} d_{in}(v) = |E| \tag{2}$$

With edge cost, the problem in Eq. (1) is NP-hard [33]. This problem is similar to the *smallest grammar problem* (SGP) [14] and in fact its NP-hardness is shown by a reduction from SGP [33].

We solve the Lexis optimization problem in Eq. (1) with a greedy heuristic, called G-LEXIS. G-LEXIS starts with the trivial flat Lexis-DAG, and at each iteration it chooses the substring ξ that maximally reduces the edge cost, when it is added as a new intermediate node to the Lexis-DAG and the corresponding edges are rewired by its addition. The algorithm terminates when there are no more substrings that reduce the cost of the Lexis-DAG. An example of application of the G-LEXIS algorithm is shown in Fig. 4. More details regarding the efficient implementation and complexity of the algorithm can be found in [33].

2.3 Path Centrality and the Core of a Lexis-DAG

After constructing a Lexis-DAG, an important question is to rank the constructed intermediate nodes in terms of significance or *centrality*. In a Lexis-DAG, a path that starts from a source and terminates at a target represents a dependency chain in which each node depends on all previous nodes in that path. Thus, the higher the number of such source-to-target paths traversing an intermediate node v is, the more important v is in terms of the number of dependency chains it participates in. More formally, let $P_D(v)$ be the number of source-to-target paths that traverse node $v \in V_M$; we refer to $P_D(v)$ as the *path centrality* of intermediate node v. Path centrality can be computed as:

$$P(v) = P_S(v) \, P_T(v) \tag{3}$$

where $P_S(v)$ is the number of paths from any source to v, and $P_T(v)$ is the number of paths from v to any target.[3] It is easy to see that $P_T(v)$ is equal to the number of times the string that corresponds to v is used in the set of targets T. Similarly, $P_S(v)$ is equal to the number of times any source node is used in the string of v, which is simply the length of that string. Hence, the path centrality of a node v is simply the product of the length of the string of v (proxy for complexity) and its number of appearances (proxy for generality).

An important follow-up question is to identify the *core* of a Lexis-DAG, i.e., a set of intermediate nodes that represent, as a whole, the most important substrings in that Lexis-DAG. The core set is the representative set of nodes that summarizes the structure of the targets. Intuitively, we expect that the core should include nodes of high path centrality, and that almost all source-to-target dependency chains of the Lexis-DAG should traverse at least one of these core nodes.

More formally, suppose K is a set of intermediate nodes and $\mathcal{P}^-(K)$ is the set of source-to-target paths after we remove the nodes in K from D. The core of D is defined as the minimum-cardinality set of intermediate nodes $Core(\tau) = \hat{K}$ such that the fraction of remaining source-to-target paths after the removal of $\hat{K}$ is at most τ[4]:

$$\hat{K} = argmin_{K \subseteq V_M} \ |K|$$
$$s.t. \ |\mathcal{P}^-(K)| \leq \tau \, |\mathcal{P}^-(\emptyset)| \tag{4}$$

where $|\mathcal{P}^-(\emptyset)|$ is the number of source-to-target paths in the original Lexis-DAG, without removing any nodes.[5] Figure 5 shows an example defining the concepts regarding the core of a Lexis-DAG.

Note that if $\tau = 0$ the core identification problem in Eq. (4) becomes equivalent to finding the min-vertex-cut of the given Lexis-DAG. In practice, a Lexis-DAG often includes some *tendril-like* source-to-target paths traversing a small number of intermediate nodes that very few other paths traverse. These paths can cause a large increase in the size of the core. For this reason, we prefer to consider the case of a positive, but potentially small, value of the threshold τ.

---

[3] A similar metric, called *stress centrality* of a vertex, is studied in [20].

[4] To simplify notation, we do not denote the core set as function of D.

[5] It is easy to see that $|\mathcal{P}^-(\emptyset)|$ is equal to the cumulative length L of all target strings.

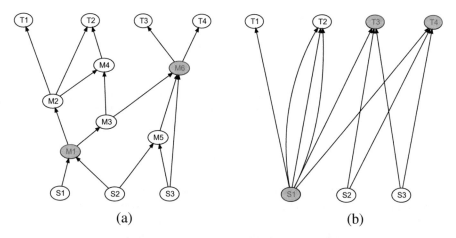

(a)   (b)

Fig. 5 (a) Original Lexis-DAG D and its core nodes highlighted (for clarity, the string of each node is not shown and the nodes are referred to with labels). For $\tau = 0.9$, we have $Core(\tau) = \{M1, M6\}$. (b) D_f, flat version of D. For same $\tau = 0.9$, we have $Core_f(\tau) = \{T3, T4, S1\}$. Hence, the H-score is $H_D(\tau) = 1 - \frac{2}{3} = 0.33$

We solve the core identification problem with a greedy algorithm referred to as G-CORE. This algorithm adds in each iteration the node with the highest path centrality value to the core set, updates the Lexis-DAG by removing that node and its edges, and recomputes the path centralities of the remaining nodes before the next iteration. The algorithm terminates when the desired fraction of source-to-target paths is achieved.

2.4 Hourglass Score

Intuitively, a Lexis-DAG exhibits the hourglass effect if it has a small core. To make this intuition more precise, we compare the size of the core of a Lexis-DAG with the core size of a derived Lexis-DAG which maintains the source–target paths of the original Lexis-DAG but that is not presenting the hourglass structure by construction.

We use a metric, named as hourglass score, or *H-score*, in our study for measuring the "hourglass-ness" of a network. This metric was originally presented in [30].

To calculate the H-score, we create a flat Lexis-DAG D_f containing the same targets as the original Lexis-DAG D. Note that D_f preserves the source–target dependencies of D: each target in D_f is constructed based on the same set of sources as in D. However, the dependency paths in D_f are direct, without forming any intermediate modules that could be reused across different targets. So, by construction, the flat Lexis-DAG D_f cannot have a non-trivial core since it does not have any intermediate nodes.

We define the H-score as follows:

$$H_D(\tau) = 1 - \frac{|Core(\tau)|}{|Core_f(\tau)|} \tag{5}$$

where $Core(\tau)$ and $Core_f(\tau)$ are the core sets of D and D_f for a given threshold τ, respectively. Note that $Core_f$ can include a combination of sources and targets, and it would never be larger than either the set of sources or targets, i.e.,

$$|Core_f(\tau)| \le min\{|S|, |T|\} \tag{6}$$

Clearly, $0 \le H(\tau) \le 1$. The H-score of D is approximately one if the core size of the original Lexis-DAG is negligible compared to the core size of the corresponding flat Lexis-DAG. Figure 5 illustrates the definition of this metric. An ideal hourglass-like Lexis-DAG would have a single intermediate node that is traversed by every single source-to-target path (i.e., $Core(1) = 1$), and a large number of sources and targets none of which originates or terminates, respectively, a large fraction of source-to-target paths (i.e., a large value of $Core_f(1)$). The H-score of this Lexis-DAG would be approximately equal to one.

3   Evo-Lexis Framework and Metrics

The Evo-Lexis framework includes a number of components that are described below. A general illustration of the framework is shown in Fig. 6.

- **Lexis-DAG:** The network that encodes the system's architecture at a given point in time. The inputs of the system are the sources of the DAG and the outputs are the targets.
- **Target Generation Model:** This model specifies the evolutionary process that creates new targets. For simplicity, we consider the addition of only new targets, not new sources. The generation of new targets can be either independent of the current hierarchy (*exogenous target generation*) or it can depend on that hierarchy (*endogenous target generation*).
- **Target Removal Model:** Models the removal of older targets. The total number of targets remains constant during the evolution of the network.
- **Hierarchy Design Algorithm:** This is how the Lexis-DAG is adjusted whenever we introduce new targets. This procedure can be as simple as building a Lexis-DAG from scratch (by running the G-LEXIS algorithm) on the set of existing targets. We refer to this approach as *clean-slate design*. On the contrary, the algorithm can be incremental, starting with the previously constructed hierarchy and incorporating new targets in a way that minimizes the adjustment cost. We refer to this algorithm an *incremental design*, and it is described next.

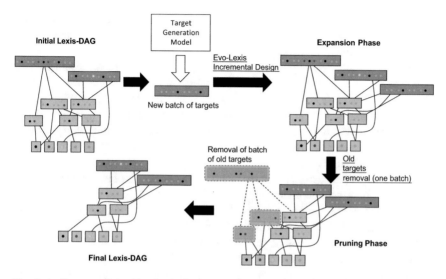

Fig. 6 A diagram of the Evo-Lexis framework. In every iteration, the following steps are performed: (1) A batch of new targets is generated via a target generation model. (2) In the "expansion phase," the new targets are added incrementally to the current Lexis-DAG by minimizing the marginal cost of adding every new target to the existing hierarchy. (3) If the number of targets that are present in the system has reached a steady-state threshold, we also remove the batch of oldest targets from the Lexis-DAG. During this "pruning phase," some intermediate nodes may also be removed because every intermediate node in a valid Lexis-DAG should have an out-degree of at least two

3.1 Incremental Design Algorithm

The Evo-Lexis algorithm generates an optimized hierarchy for the given set of targets in every evolutionary iteration. As mentioned previously, the clean-slate design approach is to discard the existing hierarchy and redesign from scratch a new Lexis-DAG for the given set of targets using the G-LEXIS algorithm. Such a design methodology is not realistic however in either technological or natural evolution. A more realistic approach is to adjust the existing Lexis-DAG incrementally, as described below.

In incremental design, given a Lexis-DAG D_0 with a set of targets T_0, a set of new targets T_+ to be added, and a set of old targets T_- to be removed, the problem is to construct a Lexis-DAG D^{INC} that supports the set of targets $\{T_0 \cup T_+ - T_-\}$, and that minimizes the cost difference with respect to D_0:

$$min_{D^{\text{INC}}} \{\mathscr{E}\left(D^{\text{INC}}\right) - \mathscr{E}(D_0)\}$$

$$\text{s.t.} D^{\text{INC}} \text{isaLexis} - \text{DAGfor}\{T_0 \cup T_+ - T_-\}$$

(7)

If $D_0 = \phi$ (i.e., there is no initial Lexis-DAG), $T_- = \phi$, and T_+ is the entire target set, the incremental design problem becomes equivalent to the original Lexis optimization problem in Eq. (1).

The incremental design problem is NP-hard (as the original Lexis design problem in which $D_0 = \phi$ and $T_- = \phi$), and so we rely on a heuristic that we refer to as INC-LEXIS. The algorithm proceeds in two phases: first, in the "expansion phase," it adds the set of new targets T_+ attempting to reuse as much as possible existing intermediate nodes. Second, in the "pruning phase," the algorithm removes the set of old targets T_-, and it also removes any intermediate nodes that are left with zero or one outgoing edges.

In more detail, the expansion phase of INC-LEXIS consists of two stages: in stage-1, we reuse intermediate nodes present in D_0 to cover T_+ with minimum cost. In stage-2 of the expansion phase, we further optimize the hierarchy that supports the targets in T_+ by building an optimized Lexis-DAG for them using G-LEXIS. The resulting new intermediate nodes and edges are added in the existing DAG.

Note that stage-1 relates to the well-known *optimal parsing* problem, which is: given a set of target strings T, a set of substrings M, and the corresponding alphabet S, what is the minimum number of substrings and letters that can construct T from the elements of $M \cup S$? The optimal parsing problem can be formulated as a shortest-path problem in directed graphs [9]. If the length of the targets is N, it can be optimally solved in $O(N + |M \cup S|)$ as the corresponding directed acyclic graph has N nodes and $O(N + |M \cup S|)$ unweighted edges.

In the pruning phase, we remove the oldest batch of targets. We also ensure that there is no redundant node in the Lexis-DAG, as implied by the constraint: $\forall v \in V_M, d_{out}(v) \geq 2$. This ensures that the Lexis-DAG does not include two types of redundancies: nodes with zero out-degree and nodes that are only reused once.

Figures 7 and 8 give an example of how INC-LEXIS adjusts a hierarchy, given a set of targets to be added and a set of targets to be removed.

3.2 Target Generation Models

The targets are generated through well-known evolutionary mechanisms, such as tinkering/mutation, recombination, and selection:

- The generation of new targets from minor changes in earlier targets is similar to *tinkering/mutation*. Tinkering is common in technological evolution: small "upgrades" in a software or hardware artifacts are the most common example of this process. In biological systems, it is well known that mutation is basically "the engine of evolution" [18]. In Evo-Lexis, tinkering/mutation is performed by replacing one character of a given target with a randomly chosen character.
- In the technological world, *recombination* is known to be one of the central mechanisms for the creation of new technologies [3]. Technological design is often considered to be a search over a space of combinatorial possibilities [40]. In fact, many breakthroughs in the history of technology were in fact just a new

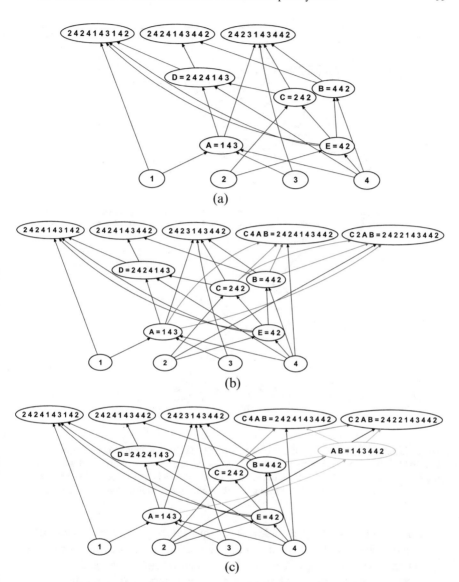

Fig. 7 Illustration of INC-LEXIS. (**a**) Initial Lexis-DAG D_0 with $T = \{2424143142, 2424143442, 2423143442\}$, $S = \{1,2,3,4\}$, and $M = \{2424143,442,242,143,42\}$. (**b**) The new targets are $T_+ = \{0424143442,2424143242,2422143442\}$. In the first stage of INC-LEXIS, the substrings in $M \cup S$ are reused to construct T_+. Red edges show the reuse of substrings in T_+. Node labels show the representation of each node using the extended alphabet formed by intermediate nodes. This representation is used in the second stage of the expansion phase to run G-LEXIS on T_+. (**c**) The Lexis-DAG after running G-LEXIS on the set T_+ in its extended alphabet form. The green nodes and edges are the results of this stage (continued in Fig. 8)

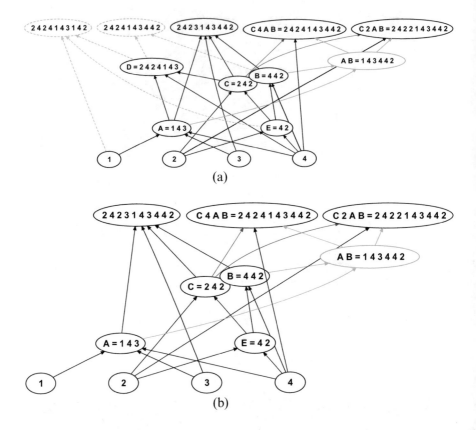

Fig. 8 (Continued from Fig. 7.) Illustration of INC-LEXIS. (**a**) The target nodes 2424143142 and 2424143442 are removed during the pruning phase. All incoming edges (dashed and shown in yellow) will also be removed, which leaves the node $D = 2424143$ with zero out-degree. (**b**) The final Lexis-DAG after removal of targets and intermediate nodes with zero and one out-degree

combination of existing modules. A recent example is the first version of the iPhone in 2007, which was introduced to be "a phone, an internet communicator and an iPod." In biology, it is well known that recombination and crossover is essential as it produces highly diverse genotypes, compared to mutations.

- *Selection* is an essential mechanism in evolution. In natural systems, selection determines whether a new genotype can survive the competition with existing genotypes (i.e., the incumbents) by evaluating the phenotypic fitness of the former relative to the latter. In the technological world, selection is the process of evaluating the functionality and cost of a new product, perhaps during an R&D cycle [31]. In the Evo-Lexis framework, selection is performed to decide whether a candidate target can be accepted, by evaluating the cost of adding that target in the current hierarchy. In other words, selection creates an *endogenous* target generation process in which the existing hierarchy determines the cost of the potential new targets and thus, whether each new target is cost-competitive compared to the targets it evolved from.

3.2.1   MRS Model

The main target generation model we consider is based on **M**utation, **R**ecombination and **S**election, thus called *MRS model*. The mechanism for this model is illustrated in Fig. 9. In detail:

1. Two distinct targets t_{s_1} and t_{s_2} (referred to as "seeds") are chosen randomly from the existing set of targets. Their cost is denoted by $C(t_{s_1}) = d_{in}(t_{s_1})$ and $C(t_{s_2}) = d_{in}(t_{s_2})$, respectively, and it is equal to the number of incoming edges that form t_s from the intermediate nodes in the current Lexis-DAG.

2. A randomly chosen "crossover index" $1 \leq i \leq k - 1$ is chosen (recall that k is the length of the targets) and the following recombinations are generated:

   - $t_1^* = t_{s_1}[1 : i - 1] + \widetilde{c} + t_{s_2}[i + 1 : k]$
   - $t_2^* = t_{s_2}[1 : i - 1] + \widetilde{c} + t_{s_1}[i + 1 : k]$
   - $t_3^* = t_{s_2}[i + 1 : k] + \widetilde{c} + t_{s_1}[1 : i - 1]$
   - $t_4^* = t_{s_1}[i + 1 : k] + \widetilde{c} + t_{s_2}[1 : i - 1]$

   where the numbers in braces show string indices, and $\widetilde{c}$ is a randomly chosen character that represents the mutated element. In other words, each recombination also includes a single-character mutation.

3. For each of the four recombinations, we calculate its cost when it is added as a new target to the current Lexis-DAG. This cost can be seen as the marginal overhead that t_x^* introduces when added to the current hierarchy D_0:

$$C(t_x^*) = \mathscr{E}\left(D^{\mathrm{INC}}(D_0, \{t_x^*\})\right) - \mathscr{E}(D_0) \qquad (8)$$

where $D^{\mathrm{INC}}(D_0, \{t_x^*\})$ is the new hierarchy after adding t_x^* to D_0 using the INC-Lexis algorithm.

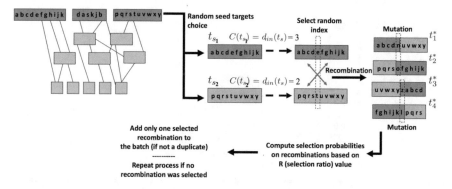

Fig. 9 Illustration of MRS model

4. The model selects a newly generated recombination t_x^* if it satisfies the following selection constraint:

   - Suppose t_x^* is formed by recombining the fragments t_{x_1} (from t_{s_1}) and t_{x_2} (from t_{s_2}), where the length of these target fragments is $|t_{x_1}|$ and $|t_{x_2}|$. The *selection ratio* is defined as:

$$R = \frac{C(t_x^*)}{|t_{x_1}| \times C(t_{s_1}) + |t_{x_2}| \times C(t_{s_2})} \tag{9}$$

   - If $R \leq 1$, we definitely accept t_x^*.
   - If $R > 1$, we accept t_x^* probabilistically with *selection probability* $p = e^{-\beta(R-1)}$.

5. If none of the recombinations passes the previous selection constraint, the target generation process is repeated. However, if one or more recombinations pass the selection constraint, the model chooses one of them randomly and adds it as an accepted target in the batch of new targets.

β determines how strongly the current hierarchy influences the selection of new targets. The larger the parameter β is, the less likely it becomes that a new target that is more costly than its seeds (i.e., $R > 1$) will be selected. For large β, we get *strong selection* and refer to the model as *MRS-strong*. A small β implies *weak selection*, and the model is referred to as *MRS-weak*. We use $\beta = 1$ and $\beta = 12$ for weak and strong selection, respectively. Figure 10 shows the difference of the two β values for typical values of R (when $R > 1$).

To analyze the effect of each evolutionary mechanism, we also consider target generation models by removing certain elements from the MRS model—hence the name "ablation study."

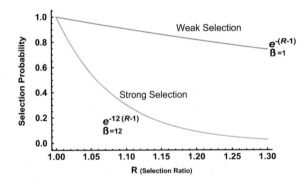

Fig. 10 The difference of the new target acceptance probability for weak ($\beta = 1$) and strong ($\beta = 12$) selection. R is the ratio between the cost of the new candidate target and the cost of the targets it evolved from. In MRS-weak, the probability of accepting the new target is high. However, this probability quickly drops in the MRS-strong model

3.2.2 MS Model

The MS model is derived from MRS by removing recombination (hence the name Mutations + Selection model or MS model). The model generates new targets as follows:

1. A target seed t_s is chosen from the existing set of targets. Suppose the cost of t_s is $C(t_s) = d_{in}(t_s)$ in the current Lexis-DAG D_0.
2. The seed is mutated (single-character mutation), as in MRS model, to t_s^*.
3. We calculate the cost of adding t_s^* to the current Lexis-DAG. This cost can be seen as the marginal overhead that t_s^* introduces when it is added to the current Lexis-DAG:

$$C(t_s^*) = \mathscr{E}\left(D^{\text{INC}}(D_0, \{t_s^*\})\right) - \mathscr{E}(D_0) \tag{10}$$

4. The model will select the newly generated target t_s^* if it satisfies the following constraint:

- $R = \frac{C(t_s^*)}{C(t_s)}$
- If $R \leq 1$, accept t_s^*.
- if $R > 1$, accept t_s^* probabilistically where selection probability $p = e^{-\beta(R-1)}$.

Otherwise, the newly generated target is rejected and the target generation repeats.

3.2.3 M Model

This is derived from the MS model by removing the selection constraint. Note that with this change the target generation process is not influenced by the current Lexis-DAG and it operates "exogenously" to the hierarchy. This model is referred to as Mutation model (or M model) and it generates targets as follows:

1. Among the targets that exist in the current Lexis-DAG, a seed target t_s is chosen randomly.
2. The seed target t_s is mutated to t_s^* through a random single-character mutation.
3. If the newly generated target t_s^* is a duplicate of one of the existing targets, the new target is rejected and the target generation repeats. If not, the generated target is added to the batch of new targets.

3.2.4 RND Model

We also consider a random target generation process, referred to as *RND*, where tinkering/mutation is removed from Mutation model. In this model, a new target is randomly generated using k random and independent choices among the sources.

3.3 Key Metrics

3.3.1 Cost Metrics

Normalized Cost This is the cost of the Lexis-DAG D_T (the Lexis-DAG for the target set T) normalized by the total length of the targets, $\mathscr{L}_T$. We denote the normalized cost by $\mathscr{C}_{\mathscr{N}}(D_T)$:

$$0 \leq \mathscr{C}_{\mathscr{N}}(D_T) = \frac{\mathscr{E}(D_T)}{\mathscr{L}_T} \leq 1 \tag{11}$$

Penalty of Incremental Design (PID) This measure evaluates the cost overhead of incremental design relative to a clean-slate design:

$$PID_T = \frac{\mathscr{E}(D_T^{\mathrm{INC}})}{\mathscr{E}(D_T^{\mathrm{CS}})} \tag{12}$$

where D_T^{INC} is the incremental design for the target set T, and D_T^{CS} is the clean-slate design for the same set of targets. The value of PID is bounded as follows:

$$1 \leq PID_T \leq \frac{\mathscr{L}_T}{\mathscr{E}(D_T^{\mathrm{CS}})} \tag{13}$$

because an incremental design cannot be more efficient than a clean-slate design (at least when the two design problems are optimally solved), and the maximum cost of incremental design is $\mathscr{L}_T$.

3.3.2 Topological Metrics

Average Depth This metric is an indicator of how deep a Lexis-DAG hierarchy is. For each target t, we calculate the average length of all source–target paths ending on that target: $\overline{d}(t)$. The average across all t is defined as the average depth of the hierarchy:

$$\overline{\mathscr{D}}(D_T) = \frac{\sum_{t \in T} \overline{d}(t)}{|T|} \tag{14}$$

Core Stability We have already defined the core size and the H-score (Sect. 2). Here we define an additional metric, related to the stability of the core across time.

   We track the stability of the core set by comparing two core sets at two different times. A direct comparison of the core sets via the Jaccard index leads to poor results. The reason is that often the strings of the two sets are similar to each other but not completely identical.

Thus, we define a generalized version of Jaccard similarity that we call *Levenshtein–Jaccard similarity*:

- The Levenshtein distance $LD(s, t)$ between two strings s and t is the number of deletions, insertions, or substitutions required to transform one string to another. The higher the number of required operations, the more distant two strings are from each other [16].
- Suppose we aim to compute the similarity of two sets A and B of strings. We define the mapping $A \rightarrow B$ where every element $a \in A$ is mapped to the most similar element $b \in B$. We also define the mapping $B \rightarrow A$ from every element $b \in B$ to the most similar element $a \in A$:

$$\begin{cases} A \rightarrow B = \{(a, b) \ s.t. \ a \in A \ \& \ b \in B \ \& \ b = arg \ max_{x \in B} Sim(a, x)\} \\ B \rightarrow A = \{(b, a) \ s.t. \ a \in A \ \& \ b \in B \ \& \ a = arg \ max_{x \in A} Sim(b, x)\} \end{cases}$$

(15)

where $Sim(a, b)$ is the similarity of a to b and is calculated as:

$$Sim(a, b) = 1 - \frac{LD(a, b)}{max(|a|, |b|)}$$

(16)

Notice that $max(|a|, |b|)$ is the maximum value of Levenshtein distance between a and b. This ensures that if $a = b$ then $Sim(a, b) = 1$, and if a and b have the maximum distance, then $Sim(a, b) = 0$.

- Considering both $A \rightarrow B$ and $B \rightarrow A$, we get the union of the two mappings and define the Levenshtein–Jaccard similarity as follows:

$$LevJac(A, B) = \frac{\sum_{(a,b) \in A \rightarrow B} Sim(a, b) + \sum_{(b,a) \in B \rightarrow A} Sim(b, a)}{(|A| + |B|)}$$

(17)

We can see that if $A = B$ (all weights are equal to one) then $LevJac(A, B) = 1$. Also if none of the elements in A are similar to B (all the element pairs take zero similarity value), then $LevJac(A, B) = 0$.

For example, suppose that $A = \{abc, cdef, fgh\}$ and $B = \{abcd, cgef, xyh\}$. The similarity of the most similar pairings is shown next:

$$\begin{cases} A \rightarrow B = \{(abc, abcd), (cdef, cgef), (fgh, xyh)\} \\ \quad where : Sim(abc, abcd) = \frac{3}{4}, Sim(cdef, cgef) = \frac{3}{4}, Sim(fgh, xyh) = \frac{1}{3} \\ \quad \Rightarrow \sum_{(a,b) \in A \rightarrow B} Sim(a, b) = 1.83 \\ B \rightarrow A = \{(abcd, abc), (cgef, cdef), (xyh, fgh)\} \\ \quad where : Sim(abcd, abc) = \frac{3}{4}, Sim(cgef, cdef) = \frac{3}{4}, Sim(xyh, fgh) = \frac{1}{3} \\ \quad \Rightarrow \sum_{(b,a) \in B \rightarrow A} Sim(b, a) = 1.83 \end{cases}$$

(18)

Hence, we have:

$$LevJac(A, B) = \frac{\sum(A \to B) + \sum(B \to A)}{|A| + |B|} = \frac{1.83 + 1.83}{3 + 3} = 0.61 \qquad (19)$$

3.3.3 Target Diversity Metric

Suppose we have a set of strings $T = \{t_1, t_2, \ldots, t_n\}$. The goal is to provide a single number that quantifies how dissimilar these elements are to each other.

- We first identify the *medoid* $\mathcal{M}_T$ within the set T, i.e., the element that has the lowest average distance from all other elements. We use Levenshtein distance:

$$\mathcal{M}_T = arg\ min_{m \in T} \sum_{t \in T} LD(t, m) \qquad (20)$$

- To compute how diverse the elements are with respect to each other, we average the distance of all elements from the medoid. We call this measure σ_T, the *diversity* of set T. The bigger the diversity metric, the more diverse the set of strings is (because the distance of each target from the medoid is the number of single-character operations needed to convert any element within the set to the medoid):

$$\sigma_T = \frac{\sum_{t \in T} LD[t, \mathcal{M}_T]}{|T|} \qquad (21)$$

4 Computational Results

4.1 Parameter Values and Evolutionary Iteration

We can summarize an evolutionary iteration of the Evo-Lexis framework as follows:

1. Initially, we start with a small number s of randomly constructed targets. Each target has the same length k, and the number of possible sources is n. An initial Lexis-DAG is constructed using the G-LEXIS algorithm.
2. In every evolutionary iteration, the following steps are performed:

   (a) A new batch of b targets is generated via a target generation model.
   (b) In the incremental design approach, the Evo-Lexis algorithm adjusts the existing hierarchy minimizing the marginal cost of adding each new target in the existing hierarchy.
   (c) If the total number of targets that are present in the system have reached a steady state (the number of targets is T_s), we also remove the oldest

Table 1 Definition and parameter values of Evo-Lexis in the following experiments

Parameter	Definition	Value
s	Number of initial targets	10
n	Number of sources	100
k	Target length (characters)	200
b	Batch size for new targets birth/old targets death	10
T_s	Steady-state number of targets present in Lexis-DAG	100

batch of b targets from the Lexis-DAG. This target removal process may also trigger the removal of intermediate nodes that are not reused by at least two other nodes in the hierarchy. The total number of targets remains constant (T_s) because the number of target additions is equal to the number of removals (b).

(d) The evolutionary process is repeated for a user-specified number of iterations. The parameters n, k, and b do not change during this process. We run each model ten times for a total of 5000 iterations. We take the mean value of each metric.

The parameters used in the following experiments are presented in Table 1.

4.2 Results

4.2.1 Emergence of Low-Cost Hierarchies Due to Tinkering/Mutation and Selection

In Fig. 11a and b, we observe a significant reduction in the normalized cost between the RND model and all other models. The main reason for this reduction is that in all other models, we generate targets that are similar to earlier targets and not randomly constructed. Further, we observe that endogenous models (MS-strong and MRS-strong) further reduce the cost of the resulting hierarchies. The reason is the large bias for selecting targets that can be constructed with lower (or comparable) cost than the seed targets they evolved from. Thus, introducing tinkering/mutation and selection both contribute to the emergence of more efficient hierarchies in the Evo-Lexis framework.

4.2.2 Low-Cost Design Resulting in Deeper Hierarchies and Reuse of More Complex Modules

Having a lower cost hierarchy also means that intermediate nodes are reused more frequently and/or that those intermediate nodes are more complex (i.e., longer strings). We observe this across models in Fig. 11c–f—models with lower normalized cost have deeper Lexis-DAGs and higher intermediate node length. These longer reused nodes further decrease the cost of the hierarchy. Hence, tinkering/mutation and selection also develop deeper hierarchies with longer intermediate nodes. These two outcomes are ubiquitously observed in both natural and technological systems. Examples include call-graphs and metabolic networks. For instance, for the OpenSSH call-graph and the monkey metabolic network, it has been reported that the underlying dependency networks have an average depth of 10.4 and 8.1, respectively [30].

4.2.3 The Recombination Mechanism Creates Target Diversity

Realistic hierarchies should support a diverse set of requirements or outputs. For example, in network protocol stacks, many different functionalities at the top level of the hierarchy (application layer) are supported by the same hierarchical infrastructure. In our framework, this translates to having a set of targets with high diversity. In Fig. 11g and h, we show the target diversity across different models. The RND model produces the highest target diversity as there are no correlations among the generated targets. In Fig. 11h, we observe that the tinkering/mutation in the M model results in 50–70% decrease in target diversity. Strong selection in the MS-strong model further decreases the diversity to the point that the targets are almost identical, with only minor variations of the same main string. Such low target diversity is not realistic in natural and technological systems. The reason that the MS-strong model behaves in this manner is that it generates new targets only through single-character mutations and only when the resulting mutants can be constructed using the existing intermediate nodes (otherwise they would have much higher cost and they would not be selected). Hence, the set of accepted new targets gets very narrow and quite similar to its seed targets.

In biological systems, the evolution of complex species required recombination and sexual reproduction (i.e., crossover). Similarly in the Evo-Lexis framework, the addition of recombination in the MRS model results in increased target diversity (Fig. 11g) while keeping the earlier properties of the Lexis-DAGs (i.e., low cost, large depth, and long intermediate nodes).

4.2.4 Reuse of Complex Modules in the Core Set by Strong Selection

Looking at the contents of the core at the 5000th iteration of all models in Fig. 13 shows that in models without selection, or with weak selection, the core includes only a small number of intermediate nodes. The reason is that random mutations make the reuse of longer intermediate nodes unlikely. Note that this does not mean that long intermediate nodes do not exist in Lexis-DAGs under the M & MS-weak & MRS-weak models—such nodes are less likely however to be reused often. As a result, shorter nodes and mostly sources are more likely to appear in the new targets, and end up in the core set.

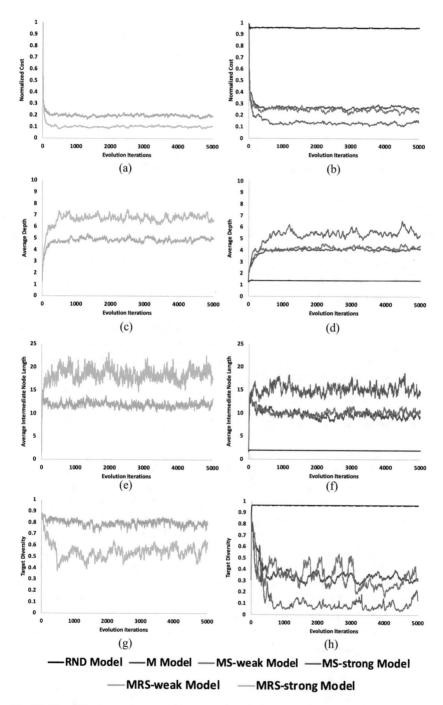

Fig. 11 Normalized cost, (average) hierarchical depth, (average) intermediate node length, and target diversity of Lexis-DAGs produced by various target generation models (weak selection models: $\beta = 1$, strong selection models: $\beta = 12$). (**a**, **b**) Normalized cost. (**c**, **d**) Average depth. (**e**, **f**) Node length. (**g**, **h**) Target diversity. (Continued in Fig. 12)

On the other hand, models with strong selection (MS and MRS) limit the locations where the seed(s) can be mutated when generating new targets. This constraint results in reusing longer intermediate nodes. Thus, selection creates a bias towards the reuse of longer intermediate nodes. In the long run, this results in some long nodes dominating the core set in the MS-strong and MRS-strong models (Fig. 13d and f).

4.2.5 Emergence of Hourglass Architecture Due to the Heavy Reuse of Complex Intermediate Modules in Models with Strong Selection

Appearance of longer and heavily reused intermediate nodes in the models with strong selection means that the architecture exhibits the hourglass effect. Indeed, we observe in Fig. 12a and b that the core size gets significantly smaller in the presence of strong selection (MS and MRS models). Additionally, Fig. 12c and d shows that the MS-strong and MRS-strong models also result in higher H-score values (0.4 and 0.65 on average, respectively). Lexis-DAGs with high H-score values have a small core size with respect to the equivalent flat Lexis-DAG whose core is made up of sources and targets only.

Overall, the reuse of longer intermediate nodes (Fig. 13) caused by selection results in hierarchies with an hourglass architecture. This observation is consistent with a mechanism (known as *reuse-preference* [30]) that was proposed earlier for the emergence of the hourglass effect in general dependency networks.

4.2.6 Stability of the Core Set Due to Selection

Selection also promotes the stability of the core set, as shown in Fig. 12h for the MS-strong model. We see an increase in core stability (i.e., similarity of the core during evolution) compared to the MS-weak and M models whose cores mostly consist of sources. Similarly, a stable core is also observed in the MRS-weak and MRS-strong models in Fig. 12g. We have already seen that long intermediate nodes appear more often in the core set of models with strong selection. Hence the core stability results show that selection not only contributes to the emergence of a small core, consisting of few highly reused intermediate nodes, but it also promotes the conservation of these core nodes during evolution. This is in agreement with the properties of several systems in which the waist of the hourglass architecture includes critical· modules of the system that are highly conserved [1, 30]. We return to this point in Sect. 7, where we further show that this core stability is occasionally interrupted by major transitions and punctuated equilibria.

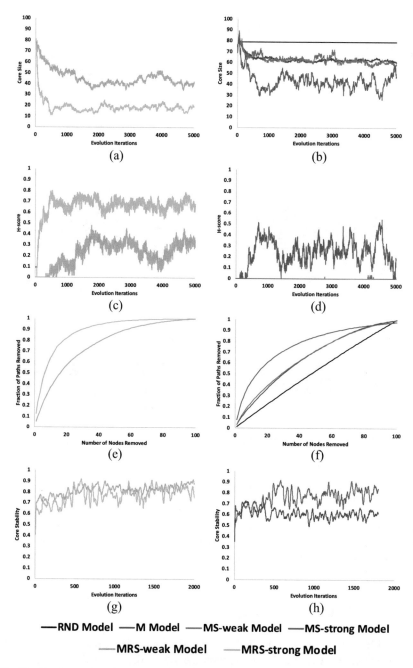

Fig. 12 (Continued from Fig. 11.). Core size, H-score, robustness to core node removals, and core stability of Lexis-DAGs produced by various target generation models (weak selection models: $\beta = 1$, strong selection models: $\beta = 12$). For core selection, we set $\tau = 0.85$. For core stability, a sliding window equal to the size of 10 batches is used to track changes in the core set. (**a, b**) Core size. (**c, d**) H-score. (**e, f**) Robustness analysis. (**g, h**) Core stability

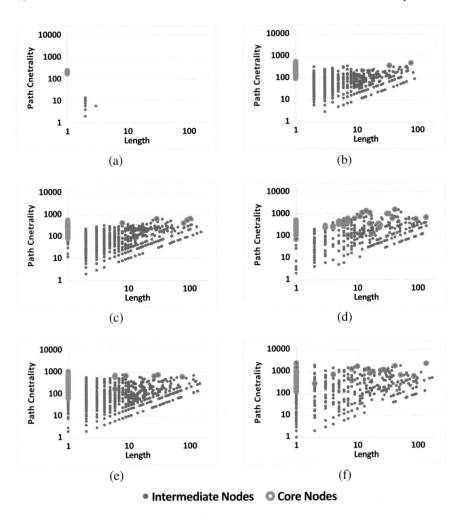

Fig. 13 Comparison of node length and path centrality in Lexis-DAGs at the 5000th iteration (for weak selection model $\beta = 1$ and for strong selection model $\beta = 12$). For core selection, we set $\tau = 0.85$. (**a**) RND model. (**b**) M model. (**c**) MS-weak. (**d**) MS-strong. (**e**) MRS-weak model. (**f**) MRS-strong model

4.2.7 Fragility Caused by Stronger Selection

Figure 12e and f shows how the generated hierarchies perform in terms of *robustness*, when we remove the most central nodes in the system, i.e., the members of the core. Robustness generally relates to the ability to maintain a certain function even when there are internal or external perturbations [30]. Figure 12f and e shows how the removal of one or more core nodes, in order of importance, contributes to cutting source–target paths in each of the Lexis-DAGs produced (at the 5000th iteration of each model).

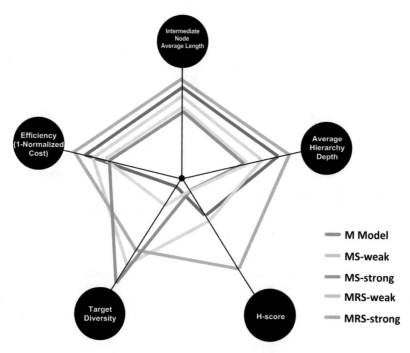

Fig. 14 Visualizing the various properties of the generated hierarchies that emerge from each model described in Sect. 5 (excluding the RND model). The MRS model produces all properties. This figure shows an approximate value for each metric at the 5000th iteration of evolution. We define, Efficiency = 1 − NormalizedCost

In hourglass architectures (MS-strong and MRS-strong model), core nodes contribute much more significantly to the overall hierarchy by covering many more source–target paths. Hence, such architectures are fragile if the core nodes are perturbed. This is similar to the concept of removal of hub nodes in scale-free network [8]. Weakening selection reduces the H-score (as in Fig. 12c) and hence, reduces the contribution of core nodes in covering source–target paths.

Figure 14 summarizes the properties of the hierarchies that emerge in the models we described in this section.

5   Evolvability and the Space of Possible Targets

As shown in the previous section, the MRS-strong model leads to hourglass hierarchies, maintaining at the same time significant target diversity. In this section, we further show that hourglass architectures have two important properties. On the positive side, they are more evolvable in the sense that new targets can be constructed at a low cost, mostly reusing the intermediate modules in the core of the hierarchy. On the negative side however, hourglass architectures only accept a

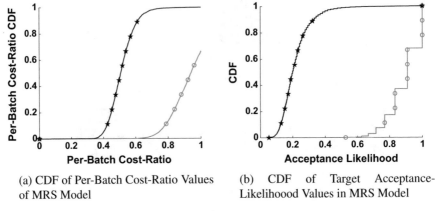

(a) CDF of Per-Batch Cost-Ratio Values of MRS Model

(b) CDF of Target Acceptance-Likelihoood Values in MRS Model

—⊖— MRS-weak Model —★— MRS-strong Model

Fig. 15 (**a**) CDF of MRS-over-MR per-batch cost-ratio, the ratio between the average cost of targets per batch in the MRS model (weak or strong selection) over the average cost of targets per batch in the MR model. (**b**) CDF of the target acceptance-likelihood, i.e., the number of accepted targets generated per batch in the MRS model divided by the total number of generated targets per batch with the same model

small fraction of the candidate new targets, restricting what a biologist would refer to as the "phenotypic space" of the system. This interplay between evolvability and the space of feasible system phenotypes or functions is an important issue in both biological and technological systems (e.g., Internet architecture [29]).

We first look at the cost of targets produced with and without selection. For this purpose, we compare two models: one is the MRS-strong model that acts as an "endogenous" target generation process. The other is a variation of MRS without selection that we call *MR model* (only mutations and recombination)—this is an "exogenous" target generation process that does not depend on the current state of the hierarchy. The MR model allows us to examine how selection affects the cost and space of acceptable targets with and without the selection constraint.

In Fig. 15, we calculate the ratio between the average cost of accepted targets per batch in the MRS-strong model over the corresponding cost in the MR model—we refer to this as *MRS-over-MR per-batch cost-ratio*. The average and median values of this ratio are 0.53 and 0.52, respectively. This suggests that the targets generated under stronger selection are of much lower cost (around half) compared to the targets generated without selection. So, the presence of strong selection allows the system to construct new targets at a much lower cost because those selected targets can be constructed mostly reusing the intermediate nodes present in the hierarchy.

As a result of strong selection, the *acceptance-likelihood* of new targets generated by the MRS-strong model is much lower than that with the MR model. Specifically, the acceptance-likelihood in Fig. 15b is defined as the fraction of accepted targets generated per batch. The mean and median of this likelihood in the MRS-strong model are equal to 0.2. In other words, about 80% of the new targets generated through mutations and recombination are not selected because their cost, given the existing architecture, would be prohibitively high.

It should be also noted that the MRS-weak model behaves quite similar to the MR baseline in terms of both the MRS-over-MR cost ratio and the target acceptance-likelihood.

Overall, the results in this section show that despite having the benefit of lower cost new targets, and thus higher evolvability, selection restricts significantly the phenotypic space of accepted new targets. Given that the MRS-strong model generates hourglass architectures, we can summarize as follows: hourglass-like hierarchies under the MRS-strong model allow the construction of new functions (accepted targets) at a low cost, by mostly reusing core modules, but at the same time such architectures significantly restrict which of these functions can be supported. Targets that are quite different than the intermediate modules of the existing hierarchy would most likely not be selected.

6 Major Transitions

Major transitions have been an important and interesting phenomenon in both natural and technological evolution. Such transitions create significant shifts in evolutionary trajectories, ecosystems, and "keystone species" [21]. There are many examples of such events in natural systems, such as the "invention" of sexual reproduction and evolution of multicellularity [34]. In technological evolution, innovations occasionally lead to the emergence of disruptive new technologies, such as the steam engine in the nineteenth century or air transportation in the twentieth century. In the context of computing, the evolution of programming languages has gone through punctuated equilibria, interrupted by new languages that were developed by tinkering or combining different structural components of older languages [37].

The results of Fig. 12g suggest that the structure of the core is locally stable, when comparing core nodes in adjacent iterations. To further investigate the stability of the core during evolution, we focus on the most central node in the core of the Lexis-DAGs, i.e., the core node that covers the largest fraction of source–target paths. We refer to this node of the Lexis-DAG as *top-1 core node*.

First, we track the variability of this node locally, by comparing its normalized Levenshtein distance to the top-1 core node in the next iteration. Figure 16 shows the results of this analysis for both MRS-strong and MRS-weak. In the MRS-strong model, we observe that in most iterations the top-1 core node does not change significantly. Even though there are some spikes in which the Levenshtein distance is larger than 0.2, in 82.6% of the evolutionary iterations the variability of the top-1 core node is less than that. Further, there are several *stasis periods* in which the top-1 core node is practically the same (Levenshtein distance lower than 0.1 or even 0). In Fig. 16 we highlight with red vertical lines a small number of stasis periods in which

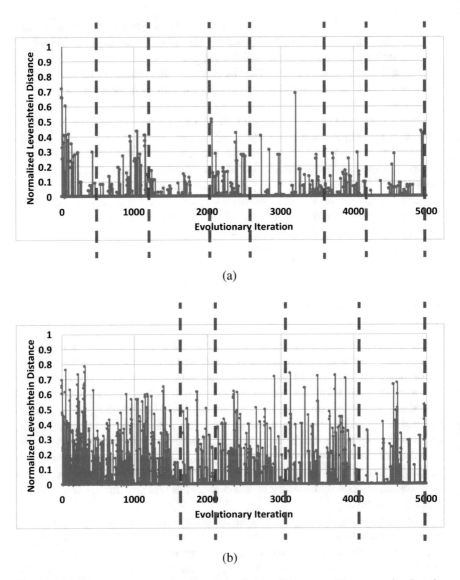

Fig. 16 Variability across successive iterations of the top-1 core node (measured using the Levenshtein distance) in the MRS model (both strong and weak selection). The highlighted iterations illustrate some of the stasis periods, in which the top-1 core node remains identical for many iterations. (**a**) MRS-strong model. (**b**) MRS-weak model

the top-1 core node remains exactly the same for tens of hundreds of iterations. On the other hand, the MRS-weak model has significantly higher variability in the top-1 core node, and fewer/shorter stasis periods. This suggests that selection is the key factor in generating these long periods of stability in the core of the hourglass architecture.

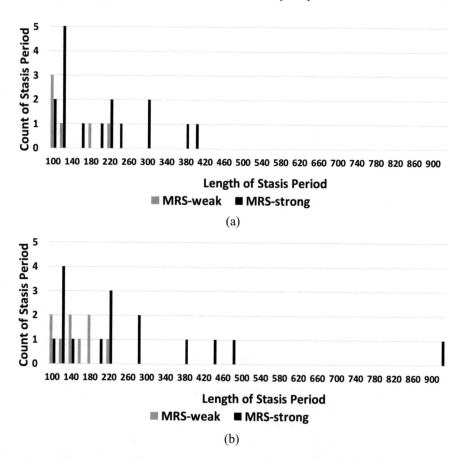

Fig. 17 Count of stasis periods (lasting at least 100 iterations) for two values of the Levenshtein distance threshold, μ_{LD}, in Fig. 16. Strong selection leads to longer and more frequent stasis periods. (a) $\mu_{LD} = 0.1$. (b) $\mu_{LD} = 0.2$

To further quantify this point, we focus on stasis periods that last at least 100 iterations (recall that the entire evolutionary paths in these results consist of 5000 iterations). Figure 17 shows that there are fewer and shorter stasis periods in MRS-weak model than in MRS-strong. The fraction of iterations that account for stasis conditions is $\frac{478}{5000} \sim 0.095$ in MRS-weak, and $\frac{2928}{5000} \sim 0.585$ in MRS-strong, when the minimum Levenshtein distance is $\mu_{LD} = 0.1$ (also $\frac{1049}{5000} \sim 0.209$ in MRS-weak and $\frac{4133}{5000} \sim 0.826$ in MRS-strong when $\mu_{LD} = 0.2$).

The presence of stasis periods under strong selection suggests that the most central intermediate nodes at the waist (or core) of the hourglass architecture can be quite stable and time-invariant. What happens however across different stasis periods? Does that stability persist across different stasis periods, or does the architecture exhibit major transitions and punctuated equilibria?

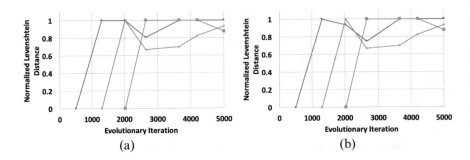

Fig. 18 Starting from three different stasis periods (with $\mu_{LD} = 0.1$), the top-1 and top-2 core node does not stay the same in subsequent stasis periods. The normalized Levenshtein distance between the top-1 and top-2 node at the start of each curve and at successive stasis periods is close to 1, suggesting that these nodes have changed. We observed similar results for other core nodes. (**a**) Top-1 core node changes in MRS-strong. (**b**) Top-2 core node changes in MRS-strong

To answer this question, we focus again on the top-1 core node and measure its variability across successive stasis periods. In Fig. 18, we consider three different stasis periods (one curve for each initial stasis period), and calculate the normalized Levenshtein distance between the top-1 core node in its initial stasis period and the top-1 core node in subsequent stasis periods. Note that the top-1 core node changes significantly across stasis periods. In fact, the Levenshtein distance is so high (often close to 1), suggesting that these are completely different core nodes. This observation gives more evidence that the top contributors to the core can lose their importance during evolutionary time scales, causing major transitions in both the core set and, consequently, in the overall hierarchy. We have confirmed that this is even more common for lower centrality core nodes too, and it is certainly even more true under weak selection.

7 Overhead of Incremental Design

In this section, we compare the cost and structural characteristics of *incremental design* (INC) relative to *clean-slate* (CS) design, i.e., the ideal case in which a new Lexis hierarchy is designed from scratch every time the set of targets is changed. Of course such clean-slate designs are rare or infeasible in practice, especially in biological evolution. CS design is still valuable however as a baseline for evaluating the cost efficiency of INC, and the hierarchy that is produced by the latter.

In the Evo-Lexis framework, a key factor that quantifies the difference between INC and CS design is the *batch size*. If the batch size b is equal to the total number of targets in steady state T_s, INC and CS are equivalent because the set of targets completely changes in each iteration. At the other extreme, if the batch size is only one target and $T_s \gg 1$, INC performs a minimal adjustment of the hierarchy to

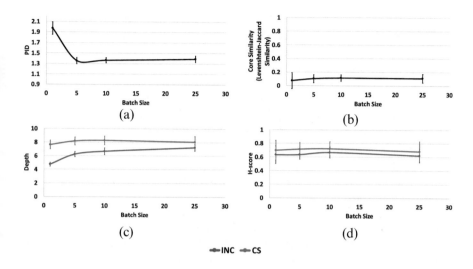

Fig. 19 Comparison between incremental (INC) design and clean-slate (CS) design, in terms of four metrics and for different batch sizes. For each batch size, the MRS-strong model is run for 5000 iterations and an average of each metric is taken over 50 distinct iterations. The considered batch sizes are: 1, 5, 10, 25. (**a**) Penalty of incremental design. (**b**) Core similarity. (**c**) Average hierarchy depth. (**d**) H-score

support the new target while CS still redesigns the complete hierarchy. In other words, the fraction b/T_s controls the degree of change in each evolutionary iteration. Both in natural and technological systems, evolution proceeds rather slowly—for this reason we only consider the lower range of this ratio, between 1/100 and 25/100.

In the following we only consider the MRS-strong model (based on the results of the earlier sections). Figure 19 compares INC and CS in terms of four key metrics. The first metric relates to cost: recall that the penalty of incremental design (PID) is the ratio of the cost of an evolving INC hierarchy over the cost of the corresponding CS hierarchy for the same set of targets. With the exception of the minimum possible batch size ($b = 1$), it is interesting that INC does *not* lead to much less efficient hierarchies than CS. The PID metric shows that INC is typically around 30% more costly than CS for a wide range of batch sizes, suggesting that INC is able to often reuse intermediate nodes in constructing the given targets, despite the fact that it cannot redesign the complete hierarchy. The PID is substantially higher when $b = 1$ however. The reason is that when the INC-Lexis algorithm is given only one new target in every iteration, it is unlikely to identify segments of that single target that repeat more than once. This means that, when $b = 1$, INC rarely adds new intermediate nodes in the hierarchy even though successive targets can be quite similar. CS, on the other hand, exploits the similarity of the set of targets in each iteration constructing more intermediate nodes, and reducing cost through their reuse.

Interestingly, even when the INC and CS designs have similar costs, they are very different in terms of the nodes that form the core. This is shown in Fig. 19b: the similarity of the two cores according to the Levenshtein-Jaccard similarity is around 0.1. This implies that the two design approaches lead to substantially different architectures in terms of the actual intermediate nodes they reuse.

Additionally, the average hierarchical depth of CS architectures is larger (see Fig. 19c) because this design approach is able to identify more and longer intermediate nodes that can be reused to construct the entire set of targets. INC, on the other hand, is constrained to not adjust the existing portion of the hierarchy, and it can only form new intermediate nodes when it detects fragments in the set of new targets that are repeated more than once. So, the INC hierarchies are typically not as deep as those in CS.

Despite their differences, both design approaches lead to hourglass architectures when the targets are created with the MRS-strong model. This is shown in Fig. 19d, and it suggests that even though INC is constrained, as described above, it is still able to identify few intermediate nodes that can be reused many times to construct the time-varying set of targets.

8 Discussion and Prior Work

The Evo-Lexis model is primarily related to three research themes: first, the emergence of modularity and hierarchy in complex systems; second, the hourglass architecture in hierarchical networks; and lastly, the comparison between offline (or "clean slate") design and online (or incremental) design.

8.1 Modularity and Hierarchy

The modeling framework of "modularly varying goals," by Kashtan and Alon, is a plausible explanation for the emergence of modularity [22, 23]. By applying incremental changes in logic circuits and evolving neural networks for pattern recognition tasks, they show that modularity in the goals (what we refer to as "targets") leads to the emergence of modularity in the organization of the system, whereas randomly varying goals do not lead to modular architectures. Similarly, Arthur et al. focus on the evolution of technology using a simple model of logic circuit gates [4]. Each designed element is a combination of simpler existing elements. Their simulation model results in a modularly organized system, in which complex functions are only possible by first creating simpler ones as building blocks. These models are similar to Evo-Lexis in the following way: when the system targets are not randomly constructed but they are generated through an evolutionary process that involves mutations, recombination, and selection, the target functions are computed through deep hierarchies that reuse common intermediate components.

Clune et al. show that modularity is a key driver for the evolvability of complex systems [15]. The authors demonstrate that selection mechanisms that minimize the cost of connections between nodes in a networked system result in a modular architecture. This is shown by evolving networks that solve pattern recognition tasks and Boolean logic tasks. The inputs sense the environment (e.g., pixels) and produce outputs in a feed-forward manner (e.g., the existence of patterns of interest). In other words, the networks that have evolved for optimizing both performance (accuracy in recognition) and cost (network connections) are more modular and evolvable (in the sense of being adaptable to new tasks) than those optimized for performance only. In a follow-up study by Mengistu et al. in [25], it is shown that the minimization of the cost of connections also promotes the evolution of hierarchy, the recursive composition of sub-modules. When not modeling the cost of connections, even for tasks with hierarchical structure (e.g., a nested Boolean function), a hierarchical structure does not emerge. These modeling frameworks are similar to Evo-Lexis because the latter also aims to minimize the number of connections in the resulting hierarchical network, and it is this cost minimization that provides the incentive for reuse of intermediate components.

At the empirical side, prior work has established that technology evolves similarly to biological evolution, through tinkering, new combinations of existing components, and selection. For instance, a study of USPTO data gives evidence for the combinatorial evolution of technology [40]. The authors find that the rate of new technological capabilities is slowing down but a huge number of combinations allows for a "practically infinite space of technological configurations." By considering technology as a combinatorial process, Kim et al. [24] uses USPTO data to investigate the extent of novelty in patents. They propose a likelihood model for assessing the novelty of combinations of patent codes. Their results show that patents are becoming more conventional (rather than novel) with occasional novel combinations.

8.2 Hourglass Architecture

A property of many hierarchical networks is the *hourglass effect*, which means that the system receives many inputs and produces many outputs through a relatively small number of intermediate modules that are critical for the operation of the entire system [30]. This property is also one of the main themes investigated in our work.

Akhshabi et al. studied the *developmental hourglass* which is the pattern of increasing morphological divergence towards earlier and later embryonic development [2]. The authors conclude that the main factor that drives the emergence of the hourglass architecture in that context is that the developmental gene regulatory networks become increasingly more specific, and thus sparser, as development progresses. Earlier, the same authors in [1] were inspired by the hourglass-resemblance of the Internet protocol stack in which the lower and higher layers tend to see frequent innovations, while the protocols at the waist of the hourglass appear to be "ossified." The authors present an abstract model, called *EvoArch*, to explain the survival of popular protocols at the waist of the protocol stack. The protocols which provide the same functionality in each layer compete with each other and, just as in [2], the increasing specificity and sparsity is what causes the network to have an hourglass architecture. The Evo-Lexis model is neither layered nor probabilistic, and so it is fundamentally different than *EvoArch*, but it also generates hierarchies in which the nodes that represent shorter strings (equivalent to lower-layer nodes in EvoArch) are reused more frequently and so they have a higher out-degree.

Friedlander et al. focus on layered networks that perform a linear input–output transformation [17] and show that in such systems the hourglass architecture emerges when that transformation is compressible. In their model, this is interpreted as rank-deficiency of the input–output matrix that describes the function of the system. A further requirement is that there should be a goal to reduce the number of connections in the network, similar to Evo-Lexis. This rank-deficiency in the input–output matrix resembles the case in which Evo-Lexis targets are not constructed independently but through an evolutionary process that generates significant correlations between different targets.

The hourglass architecture has been also investigated in general (non-layered) hierarchical dependency networks, similar to Evo-Lexis, by Sabrin and Dovrolis [30]. That analysis is based on identifying the core of a dependency network, as the minimum set of nodes that cover at least a fraction τ of all source-to-target dependency paths. We have adopted that approach, as well as the hourglass metric proposed in [30]. Their study shows the presence of the hourglass property in various technological, natural, and information systems. The authors also present a model called *reuse-preference*, capturing the bias of new modules to reuse intermediate modules of similar complexity instead of connecting directly to sources or low complexity modules.

Despite this prior work, the interplay between the emergence of hourglass architectures and cost optimization in hierarchical networks has not been explored in previous research. Evo-Lexis identifies the conditions under which the hourglass property emerges in optimized dependency networks.

8.3 Interplay of Design Adaptation and Evolution

A main theme in our study is the interplay between changes in the environment (the targets that the system has to support) and the internal architecture of the system.

Bakhshi et al. investigate a network topology design scenario in which the goal is to design a valid communication network between a set of nodes [5]. The authors formulate and compare the consequences of two different optimization scenarios for that goal: *incremental design* in which the modification cost between the two last snapshots of the design is minimized, and *optimized design* in which the total cost of the network is minimized in every increment. Focusing on the case of ring networks, even though the incremental designs are more costly, the relative cost overhead is shown to not increase as the network grows. In a follow-up study, focused on mesh networks, the same observation is made and further, the incremental design is shown to be producing larger density, lower average delay, and more robust topologies [6].

Incremental design approaches are also considered in other contexts, such as in deep neural networks (DNNs). Specifically, an important problem in machine learning is how to transfer learned features of a deep network from one task to another [39]. Transfer learning can be considered analogous to the way in which new targets are added in an Evo-Lexis hierarchy: new targets (output functions) are incrementally included in the Lexis-DAG (incrementally learned), by reusing previously constructed intermediate nodes (features of intermediate complexity) and then optimizing the part of the DAG between those nodes and the new targets (learning the weights between the existing features and the new outputs).

The incremental design policies that we consider in this paper are studied in computer science under the umbrella of *online algorithms* [32]: an online algorithm finds a sequence of solutions based on the inputs it has seen so far, without knowing the entire input sequence in advance. The main emphasis of research in online algorithms is to perform *competitive analysis*, i.e., to derive worst-case theoretical bounds between of the quality (or cost) of the solution of an online algorithm relative to its offline counterpart that knows the entire input sequence [11]. The incremental design approach in Evo-Lexis is an online algorithm but our focus is quite different: we compare empirically the cost and topological structure of the hierarchies produced by incremental design relative to an optimized ("clean-slate") algorithm that designs a minimum-cost hierarchy for the input sequence that has been seen so far.

8.4   From Abstract Modeling to Specific Evolving Systems

The Evo-Lexis model is a quite general and abstract model and it does not attempt to capture any domain-specific aspects of biological or technological evolution. As such, it makes several assumptions that can be criticized as unrealistic, such as that all targets have the same length, their length stays constant, the fitness of a sequence is strictly based on its hierarchical cost, etc. We believe that such abstract modeling is still valuable because it can provide insights about the qualitative properties of the resulting hierarchies under different target generation models. Having said that however, we also believe that the predictions of the Evo-Lexis model should be tested using real data from evolving systems in which the outputs can be well represented by sequences.

One such system is the iGEM synthetic DNAs dataset [36]. The target DNA sequences in the iGEM dataset are built from standard "BioBrick parts" (more elementary DNA sequences) that collectively form a library of synthetic DNA sequences. These sequences are submitted to the Registry of Standard Biological Parts in the annual iGEM competition. Previous research in [10, 33] has provided some evidence that these synthetic DNA sequences are designed by reusing existing components, and as such, it has a hierarchical organization. In ongoing work, we investigate how to apply the Evo-Lexis framework in the time series of iGEM sequences, and whether the resulting iGEM hierarchies exhibit the same qualitative properties we observed in this study through abstract target generation models.

9 Conclusion

We presented Evo-Lexis, an evolutionary framework for modeling the interdependency between an incrementally designed hierarchy and a time-varying set of output functions, or targets, constructed by that hierarchy. We leveraged the Lexis optimization framework, proposed in the earlier work [33], which allows the design of an optimized hierarchical network for a given set of sequences.

We developed the optimization framework, evolutionary target generation processes, and evaluation metrics needed to study the emergence and evolution of optimized hierarchies. We summarize the results of our study as follows:

1. Tinkering/mutation in the target generation process is found to be a strong initial force for the emergence of low-cost and deep hierarchies. The presence of selection, however, intensifies these properties of the emergent hierarchies.
2. Selection is also found to enhance the emergence of more complex intermediate modules in optimized hierarchies. The bias towards reuse of complex modules results in an hourglass architecture in which almost all source-to-target dependency paths traverse a small set of intermediate modules.
3. The addition of recombination in the target generation process is essential in providing target diversity in optimized hierarchies.
4. Hourglass-shaped optimized hierarchies are found to be fragile if the core nodes (i.e., nodes with highest centrality) are perturbed, similar to the concept of removal of hub nodes in scale-free networks.
5. We show that an hourglass architecture introduces a trade-off between the cost of introducing new targets and the diversity between selected targets: hourglass architectures are evolvable in the sense that they allow the introduction of new targets at a low cost but they only explore a small part of the "phenotypic space" of all possible targets. These are targets that can be constructed at a low cost reusing the larger intermediate modules in the hierarchy.
6. Our results suggest the existence of major transitions and punctuated equilibria in the evolutionary trajectory of hourglass-shaped hierarchies. The "extinction" of central modules is found to be the main factor behind this effect.
7. The comparison between incremental design and clean-slate shows that although the former is much more constrained, it has similar cost and it also exhibits the hourglass effect under the proposed evolutionary scenarios. Despite these similarities, each of these design policies results in a very different set of core modules.

Acknowledgements This research was supported by the National Science Foundation under Grant No. 1319549. We would also like to thank Matthias Gallé for his comments.

References

1. Akhshabi, S., Dovrolis, C.: The evolution of layered protocol stacks leads to an hourglass-shaped architecture. SIGCOMM Comput. Commun. Rev. **41**(4), 206–217 (2011)
2. Akhshabi, S., Sarda, S., Dovrolis, C., Yi, S.: An explanatory evo-devo model for the developmental hourglass. F1000Res. **3**, 156 (2014)
3. Arthur, W.B.: The Nature of Technology: What It is and How It Evolves. Free Press, New York (2009)
4. Arthur, W.B., Polak, W.: The evolution of technology within a simple computer model. Complexity **11**(5), 23–31 (2006)
5. Bakhshi, S., Dovrolis, C.: The price of evolution in incremental network design (the case of ring networks). In: Bio-Inspired Models of Networks, Information, and Computing Systems: 6th International ICST Conference, BIONETICS 2011, York, December 5–6, 2011, Revised Selected Papers, pp. 1–15. Springer, Berlin (2012)
6. Bakhshi, S., Dovrolis, C.: The price of evolution in incremental network design: the case of mesh networks. In: 2013 IFIP Networking Conference, pp. 1–9. IEEE, Piscataway (2013)
7. Baldwin, C.Y., Clark, K.B.: Design Rules: The Power of Modularity, vol. 1. MIT Press, Cambridge (1999)
8. Barabási, A.L., Pósfai, M.: Network Science. Cambridge University Press, Cambridge (2016)
9. Bell, T.C., Cleary, J.G., Witten, I.H.: Text Compression. Prentice-Hall, Inc., Upper Saddle River (1990)
10. Blakes, J., Raz, O., Feige, U., Bacardit, J., Widera, P., Ben-Yehezkel, T., Shapiro, E., Krasnogor, N.: Heuristic for maximizing DNA re-use in synthetic DNA library assembly. ACS Synth. Biol. **3**(8), 529–542 (2014)
11. Borodin, A., El-Yaniv, R.: Online Computation and Competitive Analysis. Cambridge University Press, New York (1998)
12. Callebaut, W., Rasskin-Gutman, D.: Modularity: Understanding the Development and Evolution of Natural Complex Systems. Vienna Series in Theoretical Biology. MIT Press, Cambridge (2005)
13. Casci, T.: Hourglass theory gets molecular approval. Nat. Rev. Genet. **12**, 76 (2010)
14. Charikar, M., Lehman, E., Liu, D., Panigrahy, R., Prabhakaran, M., Sahai, A., Shelat, A.: The smallest grammar problem. IEEE Trans. Inf. Theory **51**(7), 2554–2576 (2005)
15. Clune, J., Mouret, J., Lipson, H.: The evolutionary origins of modularity. Proc. R. Soc. Lond. B Biol. Sci. **280**(1755), 20122863 (2013)
16. Cormen, T.H., Leiserson, C.E., Rivest, R.L., Stein, C.: Introduction to Algorithms, 3rd edn. The MIT Press, Cambridge (2009)
17. Friedlander, T., Mayo, A.E., Tlusty, T., Alon, U.: Evolution of bow-tie architectures in biology. PLoS Comput. Biol. **11**(3), 1–19 (2015)
18. Hershberg, R.: Mutation–the engine of evolution: studying mutation and its role in the evolution of bacteria. Cold Spring Harb Perspect Biol **7**(9), a018077 (2015)
19. Hinton, G.E., Salakhutdinov, R.R.: Reducing the dimensionality of data with neural networks. Science **313**(5786), 504–507 (2006)
20. Ishakian, V., Erdös, D., Terzi, E., Bestavros, A.: A framework for the evaluation and management of network centrality. In: Proceedings of the 2012 SIAM International Conference on Data Mining, pp. 427–438. Society for Industrial and Applied Mathematics, Philadelphia (2012)
21. Jain, S., Krishna, S.: Large extinctions in an evolutionary model: the role of innovation and keystone species. Proc. Natl. Acad. Sci. U. S. A. **99**(4), 2055–2060 (2002)

22. Kashtan, N., Alon, U.: Spontaneous evolution of modularity and network motifs. Proc. Natl. Acad. Sci. U. S. A. **102**(39), 13773–13778 (2005)
23. Kashtan, N., Noor, E., Alon, U.: Varying environments can speed up evolution. Proc. Natl. Acad. Sci. U. S. A. **104**(34), 13711–13716 (2007)
24. Kim, D., Cerigo, D.B., Jeong, H., Youn, H.: Technological novelty profile and invention's future impact. EPJ Data Sci. **5**(1), 8 (2016)
25. Mengistu, H., Huizinga, J., Mouret, J., Clune, J.: The evolutionary origins of hierarchy. PLoS Comput. Biol. **12**(6), 1–23 (2016)
26. Miller, W. The hierarchical structure of ecosystems: connections to evolution. Evol. Educ. Outreach **1**(1), 16–24 (2008)
27. Myers, C.R.: Software systems as complex networks: structure, function, and evolvability of software collaboration graphs. Phys. Rev. E **68**, 046116 (2003)
28. Ravasz, E., Barabási, A.L.: Hierarchical organization in complex networks. Phys. Rev. E **67**, 026112 (2003)
29. Rexford, J., Dovrolis, C.: Future internet architecture: clean-slate versus evolutionary research. Commun. ACM **53**(9), 36–40 (2010)
30. Sabrin, K.M., Dovrolis, C.: The hourglass effect in hierarchical dependency networks. Netw. Sci. **5**(4), 490–528 (2017)
31. Schot, J, Geels, F.W.: Niches in evolutionary theories of technical change. J. Evol. Econ. **17**(5), 605–622 (2007)
32. Sharp, A.M.: Incremental algorithms: solving problems in a changing world. PhD thesis, Ithaca, NY (2007). AAI3276789
33. Siyari, P., Dilkina, B., Dovrolis, C.: Lexis: an optimization framework for discovering the hierarchical structure of sequential data. In: Proceedings of the 22Nd ACM SIGKDD International Conference on Knowledge Discovery and Data Mining, KDD '16, pp. 1185–1194. ACM, New York (2016)
34. Smith, J.M., Szathmary, E.: The Major Transitions in Evolution. Oxford University Press, Oxford (1997)
35. Tanaka, R., Csete, M., Doyle, J.: Highly optimised global organisation of metabolic networks. IEE Proc. Syst. Biol. **2**(4), 179–184 (2005)
36. The iGEM Web Portal. http://igem.org/main_page
37. Valverde, S., Solé, R.V.: Punctuated equilibrium in the large-scale evolution of programming languages. J. R. Soc. Interface **12**, 107 (2015)
38. Wagner, G.P., Pavlicev, M., Cheverud, J.M.: The road to modularity. Nat. Rev. Genet. **8**, 921 (2007)
39. Yosinski, J., Clune, J., Bengio, Y., Lipson, H.: How transferable are features in deep neural networks? In: Proceedings of the 27th International Conference on Neural Information Processing Systems, NIPS'14, vol. 2, pp. 3320–3328. MIT Press, Cambridge (2014)
40. Youn, H., Strumsky, D., Bettencourt, L.M.A., Lobo, J.: Invention as a combinatorial process: evidence from US patents. J. R. Soc. Interface **12**, 106 (2015)

Evaluation of Cascading Infrastructure Failures and Optimal Recovery from a Network Science Perspective

Mary Warner, Bharat Sharma, Udit Bhatia, and Auroop Ganguly

Abstract This chapter reviews the network science literature in order to create a hypothesis for the recovery of infrastructure systems. We depict the cascade of infrastructure systems through networks and simulate perturbations within a singular topology and discuss how those impact multiple layers of an interconnected network. The simulation compares and contrasts the proposed recovery methods in the literature alongside true-to-life recovery, based upon case studies throughout the United States. We explore the limitations of imposing a recovery algorithm at various points in time during infrastructure failure. This chapter aims to provide resources that account for a quantitative approach to cascading infrastructure failures, as well as accounting for human nature, politics, perceptions, and communication that may prove to be hurdles to optimizing recovery.

Keywords Critical Infrastructure · Infrastructure · Climate Science · Hazards · Resilience · Recovery · Network Science

1 Introduction

Network science has emerged as a way to study networks and better understand the world around us. With the rise of resiliency literature and a transition away from risk-based assessments alone, network science is proving to be a tool that has great potential for understanding resilience. It is uniquely positioned due to the differences that have arisen between traditional risk approaches versus risk coupled with recovery to determine resilience. Although the field of network science

M. Warner · B. Sharma · A. Ganguly (✉)
Civil and Environmental Engineering, Northeastern University, Boston, MA, USA
e-mail: a.ganguly@northeastern.edu; a.ganguly@neu.edu

U. Bhatia
Civil and Environmental Engineering, Northeastern University, Boston, MA, USA

Civil Engineering, Indian Institute of Technology (IIT) Gandhinagar, Gandhinagar, Gujarat, India

© Springer Nature Switzerland AG 2019   63
F. Ghanbarnejad et al. (eds.), *Dynamics On and Of Complex Networks III*,
Springer Proceedings in Complexity, https://doi.org/10.1007/978-3-030-14683-2_3

is relatively new as compared to many other scientific fields and tools, several approaches have emerged by which network science can be used. Infrastructure and the interdependency that critical infrastructure systems have with one another are critical areas of observation. Many communities heavily rely on critical infrastructure, which makes the resilience of infrastructure vital to their day-to-day functioning. Network science is one tool that can be used to inform decision makers on optimal strategies for increasing the resilience of specific infrastructure as well as infrastructure systems.

2 Risk and Resiliency

2.1 Assessing Risk

Traditionally, engineers and scientists have aimed to mitigate risk, which has led to a variety of approaches and frameworks to both understand and mitigate risk. A traditional approach, often referred to as probabilistic risk analysis (PRA), has been widely used as a tool to quantitatively analyze risk in industry practice and policy. One of the first studies to examine PRA came from Kaplan and Garrick [1], who chose to define risk in the form of triplets. These triplets differ throughout the literature and academic fields. However, they are consistently chosen to be three metrics for which a probability can be assessed; the intersection of these probabilities determines the level of threat [1].

The International Panel on Climate Change (IPCC) has developed a risk framework in which the intersection of three calculable metrics determines the level of threat and the resulting actions to be taken. These actions can be classified into two separate categories: (1) the calculation of risk, which is determined by the combination of vulnerability, exposure, and hazards; and (2) the resulting actions to be taken, which can be classified as adaption and/or mitigation [2]. Although the precise calculations of each input and output may be measured differently depending on the evaluator and the scope of the analysis, there are traditional ways in which these calculations are performed. Hazards are typically measured as the probability of any given hazard occurring, such as the probability of a 500-year flood hitting any given region or the probability of an earthquake of a certain magnitude hitting that same region. Vulnerability is measured in terms of value that could be lost, which may be in the form of the number of human lives lost, ecosystem pricing, or the cost of infrastructure. This particular evaluation metric is one of the most difficult because it is both great in scope but simultaneously often results in assigning a cost value in terms of dollars, which lends itself to some controversy as to whether these value assignments accurately reflect true societal values [3]. Finally, exposure is measured in terms of the probability of damage given a threat. If many exposed assets exist and a hazard event is predicted to come, these assets are not vulnerable if they will not be in the path of the hazard event, nor will they be vulnerable if they are properly secured and reinforced.

When there is a high probability of a hazard, many valuable assets, and a high probability of damage as a result of the hazard, this results in severe risk [2]. The IPCC has laid out two definitions of risk: key and emergent. This definition is typically been associated with risks related to climate hazards, but it can be broadly interpreted as well. Key risks and emergent risks are distinguished as those that are extremely time sensitive and pressing due to imminent severity (key) as opposed to those that develop over time and gradually pose an increased risk (emergent) [2]. Although time-sensitive and pressing risks are important to immediately tackle, emergent risks pose some of the greater possibilities for action and therefore must not be ignored.

Once risk is determined, actionable items usually take the form of mitigation and/or adaptation. Mitigation includes actions that defer risk or in some way help to reduce risk. Mitigation is usually in the form of deflecting the hazard event from occurring. Adaptation is the action of learning how to change the system so that it can withstand the hazard itself. The IPCC Risk Framework illustrates adaptation and mitigation separately but in conjunction with governance and socioeconomic pathways [2]. Policies and economic incentives are often used to reach the mitigation or adaptive goal, such as through cap and trade programs to reduce extreme climatic change or stricter airport screening policies to reduce manmade terror threats [2].

2.2 Gaps in the Risk Literature

The IPCC Risk Framework focuses primarily on climatic or natural events, with a particular emphasis on climate change, but it has been widely adapted as a standard for many scientists to best quantify inherent risk and response strategies [2]. However, there are still many unresolved problems that come with evaluating risk and creating responses to risk assessments, which include but are not limited to the following: risk is done through a component-wise analysis, risk is threat-specific, and risk is system-specific. Essentially, risk is confined to assessing particular areas and particular activity within those areas and emphasizes pre-event preparation.

Throughout the world in which we live, we can witness the inherent interconnectedness of most of our critical infrastructure systems. The U.S. Department of Homeland Security has identified 16 categories of critical infrastructure, ranging from the transportation sector to the food and agricultural sector, with an attempt to capture all infrastructure on which we rely [4]. When performing a risk analysis as outlined by the IPCC and others, the choice must be made between limiting interconnectedness or including all connections but thus having an overly grandiose scope in which specific implications are difficult to attribute and computational time is high. For this reason, risk analysis has a tendency to require a specific location and specific threat, and the calculations are performed component-wise in the aforementioned style of the triplets [1].

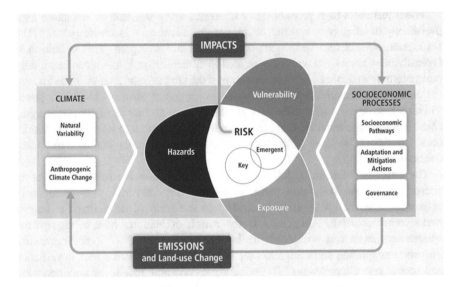

Fig. 1 A risk assessment framework that describes quantifiable ways to determine and calculate the risk of climatic changes [2]

However, this approach is problematic for several reasons. First and foremost, limiting location has action implications that may result in disjointed efforts to mitigate or adapt. For example, the probability of a hazard and its degree of impacts to Long Island, New York would also likely have impacts on downtown Manhattan (New York) due to both geographic proximity as well as shared resources and commuters between the two locations. Therefore, conducting a risk analysis and developing unique plans of action should not happen in silos. This location limitation can be scaled to greater or smaller location sizes due to the interconnected nature of all communities. There appears to be no universal solution as to where and how to draw geographic boundaries when disasters extend beyond any one confined area. Additionally, by limiting an analysis to a specific threat or hazard event, it particularly skews the public preparedness and infrastructure reinforcement. A building on the U.S. West Coast might be well-prepared for an earthquake because it is a reoccurring phenomenon, but that does not mean this same building is well-prepared for a fire, hurricane, terror attack, or cyber-attack. Each separate threat must be analyzed and acted upon separately according to a standard risk analysis (Fig. 1).

2.3 Moving Towards Resilience

Due, in part, to the growing recognition of the limitations of risk analysis, the field and concept of building and designing for resiliency has emerged. Resiliency takes the components of risk analysis and couples those results for preventative measures alongside a response to unpredictable, unforeseen, or cascading impacts. Due to the interconnectedness of our resources, infrastructure, and geographical boundaries, the theory states that it is therefore nearly impossible to accurately predict and appropriately respond to any and all risks.

Resiliency has taken on many different definitions in different contexts and particularly in different fields, with one purported estimate of 70 unique definitions [5]. The consistent underlying message is the ability to be strong and adaptive. Therefore, for the purpose of agreement and consistency, this chapter has adapted and applied resilience theory based on the definition of the National Academy of Sciences: "the ability to prepare and plan for, absorb, recover from, and more successfully adapt to adverse events" [6]. Through this definition, a risk analysis can be determined or conducted prior to an event with the intention of mitigating or adapting, as well as actively recovering when risk analysis fails to be preventative.

Nonetheless, events that have been deemed "black swan" or "grey swan" events will always exist in the literature [7]. Black swan events are extremely unlikely but cannot be said to have no probability of occurring; however, their slim likeliness makes them very unpredictable. Grey swan events are very rare but still somewhat predictable. In both such events, the impact of the hazard would be catastrophic [8]. Take, for example, the terror attacks in the United States on September 11, 2001. This was a devastating event in which little information was known or could be drawn upon to know the exact plans, timing, and approach to catastrophe. The event was essentially unpredictable. Conversely, a 500-year flood—or a series of 500-year floods—is very rare, but data do exist by which we could have some predictable power. In both scenarios, the small likelihood does not negate the large impacts. Conducting a risk analysis and determining there is low risk because of the extremely small probability does not help when such an event does occur. The definition of resilience focuses on this addendum to the standard risk analysis and measures a system or system-of-systems ability to recover once a catastrophic event has occurred. This was best illustrated by Linkov et al. [9], who illustrated perturbation impacts and optimal recovery through a curve in which functionality changes over time (Fig. 2).

A resilient system is one with very limited, or perhaps no, boundaries [9]. When conducting a risk analysis, as previously mentioned, specific areas and threats are studied and the impacts are analyzed for each component, such as a specific industry or specific infrastructure. Resiliency assumes that interconnections and interdependencies are inherent and cannot be removed. Therefore, all must be considered and any given perturbation—whether it be a natural disaster or manmade attack—should have a minimal effect on the system's ability to respond.

3 Network Science as a Tool

Because the definition of resiliency is very broad, there must be a tool to study such a broad subject area. Although many tools do exist and have been proposed, the emergence of network science has provided some of the greatest and far-reaching

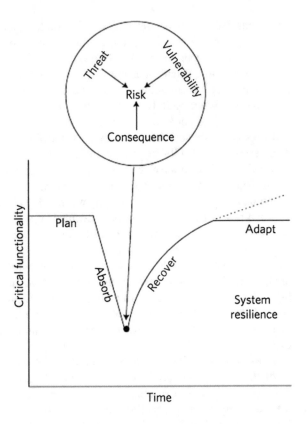

effects. Rooted in graph theory and mathematical proofs, network science is a fairly recent area of study [10, 11]. Network science, through the use of nodes and links, takes a more abstract approach to understanding resiliency: a node can have different mathematical attributes; however, to see node failure and cascading failures throughout the network, no granular detail is needed. Details are captured through mathematical applications, such as assigning weights or evaluating fitness. Impacts to any given node, as well as impacts to the system, can be analyzed in this way [10, 12].

The use of nodes and links allow for the theoretical approach to resiliency to be explored. In addition, it provides an opportunity to create multi-layer networks that represent the varying interplays between different systems. For example, although the functionality of a transportation network—such as an airline network—is reliant on the functionality of the infrastructure at each airport, it is also reliant on the communication network and communication towers. If communication were to be disabled, it would not solely impact that network but would cascade to other networks that rely upon on it [13].

Through the use of network science, various disasters can be simulated using real or hypothetical datasets. Nodes can be targeted at random, simulated terror attacks can occur, or nodes can be targeted for a certain feature. This allows the system to be threat agnostic, and a simulation of its attack can be run. Additionally, a simulation of real-time versus optimal recovery strategies can be run using network science. When system-wide prioritizations are studied and made, optimal recovery is possible because the interactions and dependencies amongst all nodes are considered.

There are also many ways in which network science can be used as a tool to understand the resiliency of critical infrastructure systems. Two common approaches are a structural approach accompanied by a dynamic approach. These two separate approaches rely on network science to evaluate resilience; however, one approach focuses on low-dimensional models to capture a relatively static (non-evolving) network, whereas the other uses system dynamics to define an inherent assigned resiliency for any given complex network. Although each approach is valuable, this chapter analyzes two case studies of seminal papers in each field to highlight, compare, and contrast the approaches [14, 15].

4 Case Studies

4.1 Studying Resilience Curves

These case studies use the methodologies and work as outlined by Bhatia et al. and Gao et al. [14, 15]. These two methodological approaches were selected because they use the theory of network science to analyze similar infrastructure systems; in addition, they serve as foundations for future work. The inherent goal and use of network science for these approaches was to make the infrastructure systems more resilient and analyze their robustness. Similarly, both papers defined resilience such that the networks are able to recover after perturbations, which is in line with the definition selected for the purpose of this chapter [14, 15].

The methodology and results from each study are described and compared using open-source data. Although neither approach in itself is uniquely superior to the other, each comes with its own caveats and contentions. As a result, we suggest different applications and uses for each approach depending upon the desired results, computational time, and data availability.

4.2 Data

For the purpose of comparison, the same data and network were used for each approach to resilience. The data for this case study were gathered from the University of Washington's "IEEE 118 Bus Test Case," which represents a portion of the American Electric Power System in the Midwestern United States per records dated December 1962. The data were saved in the IEEE Common Data Format by Rich Christie at the University of Washington in 1993. Figure 3 shows a diagram of the IEEE 118 Bus Test Case.

4.3   Limitations of the Data

As outlined by Kinney et al. [16], there are limitations to studying a power grid from a complex network analytical perspective due to the simplification of a nodes-link network in which not all essential nodes serve the same function. If all nodes necessary for functionality in the power grid are considered, then the network would have to be able to distinguish between various node functions, such as the distinction between substations and generators. In addition, the supply and flow of electricity through the system constantly varies depending on user demand and the generator from which the electricity is derived.

There are additional considerations, such as underground and above-ground systems in which unique vulnerabilities exist. The network itself is functionally and conceptually similar; however, an in-depth understanding of each node and each classification and vulnerability incurred by a node must be considered. Therefore, these data and the following model serve as a simplified proof-of-concept in order to depict the general schema of the network [16].

4.4   Network Analysis of IEEE Bus Test Case

For this analysis, it was assumed that the nodes are independent of each other and that each node serves the same function as the others. This assumption is a simplification of the network topology, but nonetheless provides general insights into the interconnectedness and interdependencies at play within the system. Figure 3 shows the network graph of the IEEE Bus Test Case [19]. The analysis was done using the NetworkX library for Python.[1] The graph has a total of 118 nodes and 179 links or edges, with the nodes representing the busses and the links representing the connections and reliance amongst them. The average degree of the graph is 3.03. The diameter—or the maximum number of link lengths from one bus to another—of the graph is 14.

Figure 4 shows a histogram of the degrees of nodes. The important insight here is that the maximum degree of the graph is 8, but the network as a whole is largely dominated by nodes of degree 2, which results in the average degree of the nodes being 3.03. This also indicates that the nodes with 7 or 8 degrees are much more

---

[1] https://networkx.github.io/.

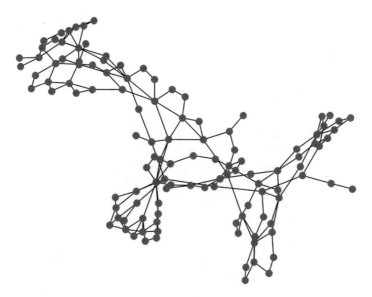

Fig. 3 The network graph

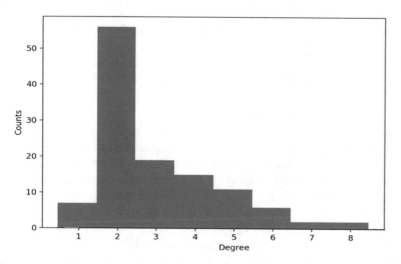

Fig. 4 Histogram of degrees

connected and potentially important to the system as a whole. This hypothesis is tested through network robustness and recovery.

4.5 Network Robustness

To check the robustness of the network, the different centralities were calculated. The centrality of the network is a measure of the importance of the nodes, such as how popular you are and/or how many people you know. Each node is assigned a centrality metric for which robustness can be determined, such that targeted or non-targeted attacks can be simulated. For example, see the following table from Newman [17]:

Centralities	Formula	Importance
Degree centrality, C^D	$C^D(i) = \frac{k_i}{N-1}$	– How popular you are – How many people you know
Betweenness centrality, C^B	$C^B(i) = \sum_{j<k} \frac{d_{jk}(i)}{d_{jk}}$	– Ability to be broken between groups – Likelihood that information originating anywhere in the network reaches you
Closeness centrality, C^C	$C^C(i) = \left[\sum_{j=1}^{N} d(i,j) \right]^{-1}$	– Being close to all nodes

Here, N is the total number of nodes, k_i is the degree of ith node, d_{jk} is the number of shortest paths between j and k, $d_{jk}(i)$ is the number of shortest paths between j and k that go through i, and $d(i,j)$ is the distance between i and j.

Based on the different measures of centrality, the nodes were removed in the decreasing order of importance to simulate a targeted attack. This is based on the assumption that a targeted attack would aim to disable the network as quickly and effectively as possible while gaining the most attention. For example, a terror attack that targets a node that is isolated or of very little importance will not gain much attention, nor will it have a great impact on the network as a whole. The random removal of a node, therefore, represents a non-targeted attack or one in which a natural disaster or an unintentional failure occurred [18].

At every time step after a node was removed, the number of nodes in the largest connected component (referred as the giant component) was measured. For example, after the first targeted attack, everything connected to the removed node would split off into several different large interconnected components. The new giant component is the one in which the most number of nodes are still connected [18].

From the graph depicted in Fig. 5, it is clear that removal of the first few important nodes (under a targeted attack) impacts the robustness of the graph substantially.

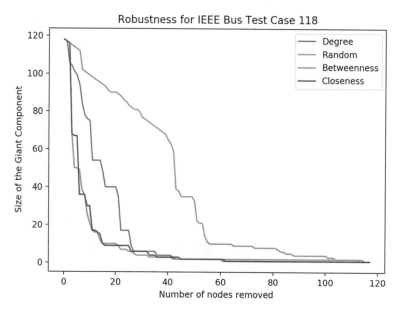

Fig. 5 Robustness graph for the IEEE Bus Test Case 118

It can be seen that an attack based on the betweenness centrality and closeness centrality affects the network the most; in this case, the targeting of the most connected node (highest degree) has slightly less impact. This procedure helps to identify the most critical nodes and how their removal affects the functionality of the network. Additionally, this process serves as a way in which decision makers can aim to strengthen the resilience of the system as a whole by knowing which nodes are of greatest threat to disabling the system.

4.6 Network Recovery

After any attack, whether it be targeted or random, it is of critical importance to restore the functionality of the system. Figure 6 shows a case of the random failure (left) that results in a complete loss of system functionality, followed by the recovery path (right) to restore the network's functionality after failure. To generate the resilience profile of the IEEE Bus Test case, the recovery algorithms from the Recovery-Master git repository[2] were used [14].

This approach has been outlined and proposed elsewhere [14]. It assumes that the network remains relatively static and the dynamics of the network do not change throughout the analysis. This assumption can be challenged, but it is also

---

[2]https://github.com/udit1408/Recovery_algorithm.

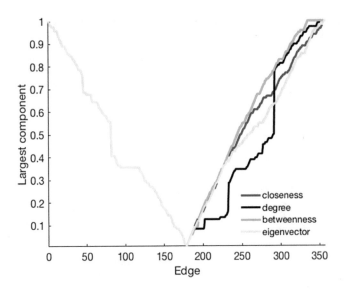

Fig. 6 IEEE Bus Test Case resilience curves

important to consider the time period over which loss of functionality occurs. Loss of functionality in the electricity grid network during an unprecedented storm may be very swift, with little action that is able to be taken until the storm passes. However, if the same analysis was conducted for the loss of biodiversity over the course of hundreds of years, then there is ample time for the structure of the network to change and adapt. For the purpose of this failure and recovery system, it remains logical to maintain the underlying assumptions.

Additionally, the conclusions from this methodology indicate to decision makers the best way in which to respond to the failure of a system. When all nodes are disjointed, no longer connected, and therefore no longer able to provide any functionality, the way in which the system recovers is important in order to return it most efficiently to full functionality. If many different operators attempt to return each individual node in the network to functionality without a targeted plan for resource allocation, then more time will be wasted. In this scenario, as outlined in Fig. 6, it is optimal to prioritize and allocate resources first towards nodes with the highest betweenness centrality, because this results in the quickest return to functionality.

4.7   Universal Resilience Curves [15]

The research done by Gao et al. [15] is unique from that of Bhatia et al. [14] in two major ways. First and foremost, Gao et al. [15] had a goal of determining how resilient any given network is, whereas the purpose of the research from Bhatia et al. [14] was to determine how best to recover a system after failure. Both studies lend insights to one another and are not mutually exclusive from one another. Additionally, the inputs and results from Gao et al. [15] take the dynamics of the system into consideration, as opposed to performing an analysis primarily on the basis of a singular topology. This requires more inputs, such as various equations regarding how to calculate the resilience, as well as a thorough understanding of the way in which the observed network acts, changes, and adapts.

The following research was conducted through the Gao et al. [15] study on measuring universal resilience; the processes and a brief understanding of the inputs are described. The authors were able to provide several equations and dynamics of the data, which produced results that demonstrated how any given network could respond to perturbation. These equations are highlighted and described as follows:

- Transmission lines have a characteristic admittance, Y. To calculate current (I) through each bus using Y and voltage at bus, V:

$$I_i = Y_{ii} V_i - \sum_{\substack{k=1 \\ k \neq i}}^{N} Y_{ik} V_k, \tag{1}$$

- To calculate load at each bus, S:

$$S_i^* = V_i^* I_i, \tag{2}$$

- To obtain a measure of system resilience, substitute Eq. (1) into Eq. (2), then multiply both sides of equation by the complex conjugate and the quadratic equation for voltage on each bus:

$$|V_i|^4 - \left(\frac{2\,\mathrm{Re}\,(S_i Y_{ii})}{|Y_{ii}|^2} + \left| \frac{1}{Y_{ii}} \sum_{\substack{k=1 \\ k \neq i}} Y_{ik} V_k \right|^2 \right) |V_i|^2 + \frac{|S_i|^2}{|Y_{ii}|^2} = 0 \tag{3}$$

Therefore, Gao et al. [15] concluded that if there is a real solution for V_{i2}, the system is functioning. If not, then the only possible outcome is blackout state with zero voltage and zero load. A nonzero solution exists if the discriminant is greater than zero:

$$\sqrt{\frac{|Y_{ii}|^2}{|S_i|\,|Y_{ii}| - \mathrm{Re}\,(S_i Y_{ii})}} \left| \frac{1}{Y_{ii}} \sum_{\substack{k=1 \\ k \neq i}} Y_{ik} V_k \right| - \sqrt{2} > 0, \qquad (4)$$

To obtain the simulations of the IEEE Bus Test Case, the network has to satisfy Eq. (4), then apply perturbation (λ) at a single node (m), such that load at node m is increased:

$$S_m \rightarrow (1 + \lambda)\, S_m.$$

Then, new activity is calculated at each node. λ is increased iteratively until the system fails at the critical perturbation, λ_c.

The simulation output is as follows:

1. For each load bus, a $5 \times M$ matrix is used, with M being the number of steps to λ_c.
2. Variables returned are voltage amplitude, β effective, lambda, x effective, and gamma.

The variable β was defined by Gao et al. [15] as representing the changing environmental conditions and x effective represents the most effective state of the system. Therefore, β effective is the conditions under which we create x effective.

In power systems, λ_c is highly dependent on the selection of the perturbed node m. For some nodes, a load increase of $\lambda \sim 1$ leads to collapse, whereas for others the system maintains its resilience even up to $\lambda \sim 10^2$—a discrepancy of two orders of magnitude. This diversity exposes the difficulty in predicting a power system breakdown [15].

Figures 7 and 8 were derived from running the data, equations, and code provided by the study via open source documentation. Figure 7 shows the performance of the power supply network under increasing demand. λ was increased until the system reached collapse. λ_c is highly unpredictable. Figure 8 shows that mapping to β-space led to much more predictable behavior, exposing the universality in power system resilience and showing that β effective captures the natural control parameter also for power supply systems. Both of the figures were created using the NuRsE git repository.[3]

---

[3]https://github.com/jianxigao/NuRsE.

Fig. 7 Performance of the
IEEE Bus Test Case under
increasing demand

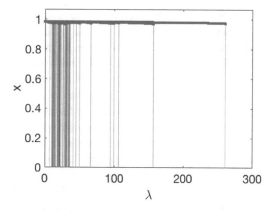

Fig. 8 Predictable behavior
when mapping to β-space

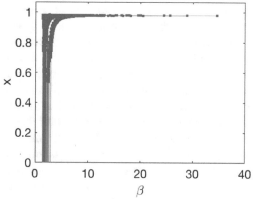

4.8   Insights and Conclusions

Based upon these two approaches, it is evident that network science serves as a
valuable tool to inform the recovery of a system. Although traditional approaches
have focused on risk assessment and methods to avoid collapse, these risk analyses
do not typically serve to fully protect a system from failure when faced with extreme
or catastrophic events. In addition, through this research, it is clear that failure of
even one node can quickly cascade through the system.

We are unable to predict all events; therefore, even the best protections and risk
analyses cannot and do not ensure no-failure. Bhatia et al. [14] focused on a way
to most efficiently recover a system once it has failed. Although there are many
underlying assumptions in this case study, including the data and network structure,
the fundamental assumption is that the need for dynamic responses and shifts in the
system is unnecessary in this scenario. There is reason to believe that this analysis
also works well for similar infrastructure systems, particularly those faced with
relatively immediate threats and destruction.

Although there is potential for degradation and dynamic shifts in infrastructure that may allow for gradual adaptation, this chapter focuses on the risks that have small probabilities of occurrence but large-scale impacts. These large impacts imply that failure would quickly cascade, although this is not inherently true.

Simultaneously, every system has hidden universal patterns of resilience [15]. These dynamics could and should be considered for an optimal understanding of a system's ability to adapt and respond to perturbation. The origin of this universality is the initial separation of the system's dynamics and topology. Gao et al. [15] also suggested potential intervention strategies to avoid the loss of resilience or design principles for optimal resilience in systems that would be able to cope with perturbations. This unique approach requires more computational time due to the necessity of knowing the dynamics of the system to be studied, and the β effective must be calculable. Although the dynamics and intricacies of many networks are known, this does not hold true for all networks. Finding the unique input values would require more data than is required by the approach in Bhatia et al. [14].

Therefore, it is safe to conclude that both approaches are of value. However, the way in which each can be used depends on the desired outcome and the inputs that would be required. Additionally, this recovery strategy [14] can be coupled with the universal resilience patterns results [15] to observe the resilience parameters of any given system, and then to study how to recover the system if it were to fail. One approach indicates how to recover, with the end results providing insight into the resilience of the network [14], whereas the other approach indicates the resilience of the network, with the end results providing insight into how to recover. These two approaches are therefore not mutually exclusive and can be coupled. The research here highlights failures in a power grid network, but it can be applied to many other networks within critical infrastructure and beyond.

References

1. Kaplan, S., Garrick, B.J.: On the quantitative definition of risk. Risk Anal. 1, 11–27 (1981). https://doi.org/10.1111/j.1539-6924.1981.tb01350.x
2. International Panel on Climate Change Secretariat: AR5 Climate Change 2014: Impacts, Adaptation, and Vulnerability. International Panel on Climate Change, Geneva (2014). https://www.ipcc.ch/report/ar5/wg2/
3. Gomez-Baggethun, E., Ruiz-Perez, M.: Economic valuation and commodification of ecosystem services. Prog Phys Geogr. 35(5), 613–628 (2011)
4. Department of Homeland Security (DHS): Critical Infrastructure Sectors. United States Cyber Security and Infrastructure Security Agency. (2018). https://www.dhs.gov/cisa/critical-infrastructure-sectors
5. Fisher, L.: Disaster responses: more than 70 ways to show resilience. Nature. 518, 35–35 (2015). https://doi.org/10.1038/518035a
6. National Research Council: Disaster Resilience: A National Imperative. The National Academies Press, Washington (2012). https://doi.org/10.17226/13457
7. Stein, J.L., Stein, S.: Gray swans: comparison of natural and financial hazard assessment and mitigation. Nat. Hazards. 72, 1279–1297 (2014)

8. Lin, N., Emanuel, K.: Grey swan tropical cyclones. Nat. Clim. Chang. **6**, 106–111 (2016). https://doi.org/10.1038/nclimate2777
9. Linkov, I., Bridges, T., Creutzig, F., et al.: Changing the resilience paradigm. Nat. Clim. Chang. **4**, 407–409 (2014)
10. Barabasi, A., Albert, R.: Emergence of scaling in random networks. Science. **286**(5439), 509–512 (1999)
11. Chawla, N., Ganguly, A.: Complex networks as a unified framework for descriptive analysis and predictive modeling in climate science. Stat Anal Data Min. **4**(5), 497–511 (2011)
12. Watts, D.J., Strogatz, S.H.: Collective dynamics of 'small-world' networks. Nature. **393**(6684), 440–442 (1998)
13. Clark, K., Bhatia, U., Kodra, E., Ganguly, A.R.: Resilience of the US National Airspace system airport network. IEEE Trans. Intell. Transp. Syst. **99**, 1–10 (2018). Accepted (minor changes)
14. Bhatia, U., Kumar, D., Kodra, E., Ganguly, A.R.: Network science based quantification of resilience demonstrated on the Indian Railways Network. PLoS One. **10**(11), 1–17 (2015). https://doi.org/10.1371/journal.pone.0141890
15. Gao, J., Barzel, B., Barabási, A.-L.: Universal Resilience Patterns in Complex Networks. Nature. **530**(7590), 307–312 (2016). https://doi.org/10.1038/nature16948. Nature Publishing Group
16. Kinney, R., Crucitti, P., Albert, R., Latora, V.: Modeling cascading failures in the North American power grid. Eur Phys J B. **46**(1), 101–107 (2005). https://doi.org/10.1140/epjb/e2005-00237-9
17. Newman, M.E.J.: Networks: An Introduction. Oxford University Press, Oxford (2010)
18. Barabasi, A., Posfai, M.: Network Science, 1st edn. University Press (2016). ISBN-10 1107076266
19. Christie, R.: Power systems test case archive (2000). http://www.ee.washington.edu/research/pstca/

Part II
Network Dynamics

Automatic Discovery of Families
of Network Generative Processes

Telmo Menezes and Camille Roth

Abstract Designing plausible network models typically requires scholars to form a priori intuitions on the key drivers of network formation. Oftentimes, these intuitions are supported by the statistical estimation of a selection of network evolution processes which will form the basis of the model to be developed. Machine learning techniques have lately been introduced to assist the automatic discovery of generative models. These approaches may more broadly be described as "symbolic regression," where fundamental network dynamic functions, rather than just parameters, are evolved through genetic programming. This chapter first aims at reviewing the principles, efforts, and the emerging literature in this direction, which is very much aligned with the idea of creating artificial scientists. Our contribution then aims more specifically at building upon an approach recently developed by us (Menezes and Roth, Sci Rep 4:6284, 2014) in order to demonstrate the existence of families of networks that may be described by similar generative processes. In other words, symbolic regression may be used to group networks according to their inferred genotype (in terms of generative processes) rather than their observed phenotype (in terms of statistical/topological features). Our empirical case is based on an original dataset of 238 anonymized ego-centered networks of Facebook friends, further yielding insights on the formation of sociability networks.

Keywords Network generators · Social networks · Machine learning · Genetic programming · Network families · Symbolic regression

T. Menezes (✉)
Centre Marc Bloch (UMIFRE CNRS/MAE), Computational Social Science Team, Berlin, Germany
e-mail: menezes@cmb.hu-berlin.de

C. Roth
CNRS, Paris, France

Centre Marc Bloch (UMIFRE CNRS/MAE), Computational Social Science Team, Berlin, Germany
e-mail: roth@cmb.hu-berlin.de

© Springer Nature Switzerland AG 2019   83
F. Ghanbarnejad et al. (eds.), *Dynamics On and Of Complex Networks III*,
Springer Proceedings in Complexity, https://doi.org/10.1007/978-3-030-14683-2_4

1   Introduction

Networks have become over the last decades a key notion for modeling systems in a wide variety of fields. This is especially so in social sciences where networks are being introduced in an increasing number of contexts. On one hand, they are a type of abstraction that lends itself very naturally to the representation of a great variety of social structures and interactions. On the other hand, the information technology revolution has been making networks both more explicitly present—for example, due to the popularity of online social media—and easy to retrieve by researchers.

Being practitioners in the field of *computational social sciences*, we have been concerning ourselves for some years with the challenge of deriving explanatory models from such complex empirical data. Networks are typically generated by phenomena that are non-linear in nature. The complex interactions between actors and the emergent environment they create—represented by the network itself— make it difficult to employ *divide-and-conquer* approaches, where the problem can be divided into smaller parts that become tractable for human researchers to reason about. In other words, it is not easy to *intuit* network formation principles which translate into simple yet successful generative models. Our belief that it makes sense to recruit computational intelligence to overcome this challenge led us to develop a method to automatically propose plausible and understandable network generators—mathematical expressions that describe how new links are formed in the network, using only local variables (e.g., the current degrees of the pair of nodes in a candidate connection). This is akin to a multi-agent system, sufficiently abstract to lend itself to the description of a variety of phenomena. In the article where we proposed the full method for the first time [73], we showed that it could be used to discover plausible and simple generators for not only social, but also biological and man-made networks.

In the last years, machine learning has been gaining popularity as a scientific tool among many fields, partly because of the recent successes in *deep learning*. We use a different approach, coming from the artificial intelligence branch usually known as *evolutionary computation*. More specifically, we use a *genetic programming* approach, given that we are evolving computer programs. There are two main reasons for this choice: the nature of the problem and the goal of understandability.

Many machine learning approaches, including the training of neural networks through back-propagation, require that an optimization criterion can be represented as a convex function, for which an optimum can be found through some form of gradient descent. The space of possible network generators appears too complex for such a convex function to be defined. In this kind of situation, evolutionary computation provides a stochastic and heuristic-driven approach to find viable solutions. The term "evolutionary" comes from its inspiration in Darwinian evolution. The simple principle of preference for the propagation of the most promising individuals with random mutations unleashes a type of intelligence that, although not human-like, is distinctly creative. To illustrate this, we show in Fig. 1 an antenna created by NASA that was designed by an evolutionary computation algorithm, aiming at optimizing

its radiation pattern. We were interested in this ability to effectively explore a complex search space while being able to entertain *counter-intuitive* solutions.

Another problem with many approaches such as neural networks is that they tend to be black boxes. Even solving the convexity problem, they might produce good results in replicating network morphogenesis, yet they do not lend themselves to creating interpretable processes. We defined our genetic programs in a simple way, and included in our method a preference for simpler programs. As we will see, they can be translated into human-readable mathematical expressions. Our results are thus comparable with classical models of network morphogenesis, for which (human) scientists are, however, usually in charge of proposing plausible formation processes.

In this chapter we provide a wider view of our work, and also share new results. In the next section we discuss the last decades of research on the modeling of network morphogenesis, while providing a systematization that aims to help situate our work within it. We pay special attention to the recent history of evolutionary models, of which we were not the only pioneers.

In Sect. 3 we provide a synthetic description of our method of symbolic regression of network generators. For all the details, we invite the reader to refer to our original article.

In Sect. 4 we present the results of novel research, aiming at finding families of generators within a dataset of networks of the same nature—in this case, ego-centered friendship networks extracted from Facebook. We were interested in finding if symbolic regression would lead to sets of similar explanations. In other words, while network families are traditionally based on phenotypical resemblance [see, e.g., 12, 31, 35, 47, 75, 79], we show here that our approach can yield families of generative processes at the level of genotype resemblance. We propose a new way to measure generator similarity, allowing us to project all the generators into a two-dimensional embedding, where generators with similar behaviors tend to be closer. With the help of this embedding, we were able to identify general patterns that many of the generator expressions conform to. From a sociological perspective, we thus also shed light on a variety of plausible mechanisms of formation of ego-centered friendship networks. More broadly, the existence of generator families further validates the behaviors embedded in the general mathematical expression characterizing a given family since it is able to efficiently reproduce the shape of several empirical networks.

2 Network Morphogenesis

To illustrate the complexity of the task of intuiting efficient generative principles, we shall first review the existing efforts in this area. We thereby intend to show better where our approach may fit in and benefit this state of the art. This will enable us to emphasize the interface position occupied by our work, which aims at inferring formation processes from the network while at the same time reconstructing it, using evolutionary modeling to avoid positing prior assumptions on the shape of these processes.

The modeling of network morphogenesis has generated a substantial literature over the last decades, especially after the early 2000s when most real-world networks were shown to exhibit peculiar connectivity and modular features. The corresponding state of the art may essentially be organized according to two key dichotomies: the first one relates to the *target* of models, the second one to their *foundations*. More precisely, (1) models aim at reconstructing either network evolution processes or morphology; and to that end, (2) they rely on assumptions, or input, related either to processes or to morphology. This yields the straightforward double dichotomy shown in Table 1, which includes a few canonical examples. Let us start by reviewing each category of that dichotomy.

Table 1 Double dichotomy of canonical network modeling approaches, which generally aim at reconstructing either evolution processes or network structure, and do so by relying either on evolution processes or network structure

	Reconstructing	
Using	Processes	Structure
Processes	Preferential attachment estimation, link prediction, classifiers, scoring methods, etc.	Preferential attachment-based generative models, rewiring, cost optimization, social simulation, agent-based models (ABMs), etc.
Structure	Exponential Random Graph Models (ERGMs), p_1, p^*, Markov graphs, stochastic actor-oriented models (SAOMs), etc.	Prescribed structure, subgraph-based constraints, Kronecker graphs, edge swaps, etc.

2.1 Reconstructing Processes

We first focus on the understanding of the generative processes at the lowest level, i.e., the rules governing the appearance or disappearance of nodes, and/or the formation or disruption of links.

2.1.1 Using Micro-Level Processes

One of the most straightforward approaches to derive these rules consists in using, precisely, data describing these very dynamics at the node and link level. In this category, we find simple counting methods aimed at appraising the propensity of links to form preferentially more towards nodes possessing certain properties—this is the archetypal notion of "preferential attachment" (PA). In its most restrictive yet most widespread acceptation, PA relates to the ubiquitous observation that links tend to attach to nodes proportionally to their degree. Following [32], this acceptation essentially stems from [16] and [56]. Several authors extended this notion beyond degrees to deal with a variety of both structural and non-structural features, including spatial distance [101], common acquaintances or topological distance [59], similarity [61, 70, 86], or a combination thereof [28]. A more recent stream of research took this approach the other way around by proposing normative growth process and comparing them with empirical link formation. For one, [80] introduced a model of link creation based on a concept of geometric optimization: nodes are placed in a plane and new nodes may connect to a subset of existing nodes by minimizing a geometric quantity. The model thereby reproduces connection probabilities observed in a selection of real networks, rather than observing real data to infer connection probabilities.

Approaches inspired by machine learning have also been proposed to abstract processes by observing processes. They principally aim at predicting the appearance of links by generalizing from past link creation. This stream is rather geared towards prediction success rather than behavior estimation, i.e., efficiently guessing which links will appear rather than providing explicit link formation rules [see 99, for a discussion of the relative performance of these methods]. Scoring methods are among the simplest of these approaches: [65] first introduced a predictor function based on some dyadic feature (such as the number of common neighbors, Jaccard coefficients, and Katz' distance). This function produces a ranking on non-connected dyads from the observation of an empirical network formed over the learning period $[t_0, t]$. The prediction task then consists in going through the dyad list in descending order and comparing it with the links that empirically appeared during a test period $[t, t']$.

A large array of more sophisticated techniques have been used in this field, by involving, inter alia, SVM classifiers [e.g., as proposed by 2], or more broadly supervised learning methods [5], as well as matrix and tensor factorization [1] (see [67] and [6] for introductory reviews of this type of endeavors). Some authors divide the network into modules, or blocks, in order to estimate a simple (and local) probability of link formation within and between these modules, e.g., [45] who define modules through stochastic blockmodeling [8], or [27] who use a dendrogram to both build the module partition and compute the inter-module connection probabilities. Overall, there has been an increasing attention to the time-related and spatial variability of the prediction task by considering the local neighborhood of nodes, both in a topological and temporal manner [89] and in a semantic fashion (e.g., by enriching the set of prediction features with content [88] or so-called sentiment analysis [102]). Also of note is the recent addition of evolutionary algorithms to this toolbox: for instance, [19] evolve a weight matrix describing the relative contributions of various similarity measures in predicting new connections.

2.1.2   Using Macro-Level Structure

Link formation principles may also be inferred from the observed network topology. The most common approach in this stream comes to econometric techniques aimed at fitting a model whose parameters are associated with specific link formation effects and which takes the whole network as an input.

Exponential Random Graph Models (ERGMs) famously belong to this class. In all generality, they rely on the assumption that the observed network has been randomly drawn from a distribution of graphs. The probability of appearance of a given graph is construed as a parameterization on a choice of typical network formation processes: be they structural (such as transitivity, reciprocity, and balance) or non-structural (such as gender dissimilarity and homophily). The aim is generally to find parameters maximizing the likelihood of the observed network. Each parameter then describes the likely contribution of the corresponding category of link formation process (e.g., strong transitivity, weak reciprocity). ERGMs have been introduced by [52] through the so-called p_1 model describing the probability of graph G as $p_1(G) \sim \exp(\sum_i \lambda_i v_i(G)) = \Pi_i \exp(\lambda_i v_i(G))$ where $v_i(G)$ denotes a value related to the i-th process (e.g., transitivity) and the λ_i are the above-evoked parameters. p_1 assumes independence between dyads, which limits the model to simple dyad-centric observables: principally, degree and reciprocity. It can nonetheless be applied to a partition of the network into subgroups [37] or stochastic blockmodels [8, 53], which posits a block structure, i.e., the fact that distinct groups of actors, or "blocks," exhibit distinct connection behaviors; parameters are thus a function of blocks. [39] later introduced "Markov graphs," which takes into account dependences between edges and thus triads and simple star structures, and which was subsequently extended as the p^* model [9, 85, 97]. Further generalizations to more complex graph structures have lately been proposed, e.g., for the so-called multi-level networks [23, 95], which are essentially graphs with two types of nodes and three possible types of links (two intra-type and one inter-type).

When longitudinal data is available, network evolution may be construed as a stochastic process. Holland and Leinhardt [51] then Wasserman [96] proposed to appraise network dynamics as a (continuous-time) Markov chain. They assumed that the probability of link appearance or disappearance depends on a limited set of (static) parameters representing the contribution of various structural effects, such as, again, reciprocity, degree. Networks observed at different points in time are used to fit these parameters. Albeit not directly affiliated with this framework, the approach of [82] proceeds in a similar fashion to determine the key factors guiding attachment of firms in a biotech sector. Stochastic actor-oriented models (SAOMs) further extend these ideas by introducing an actor-level viewpoint whereby actors establish link to optimize some objective function [91]. Again, the parameters of this function denote effects deemed important for link formation (or destruction). These models also accommodate for some form of dyadic dependence, and take into account non-structural features (including gender). They may include behavioral observables [92] or rely further on machine learning techniques, e.g., by extending SAOMs to a Bayesian inference scheme [58]. In practice, SAOMs may be used to study non-structural effects linked to gender, racial, socioeconomic, or geographical homophily, as demonstrated, for instance, in an online context on Facebook friendship [64]. ERGMs and SAOMs assuredly share several traits, and it is also possible to develop ERGMs in a longitudinal framework as temporal ERGMs (or TERGMs), where the estimation for a graph at time t depends on the graph at $t - 1$ [48]. For a more detailed comparison between SAOMs and ERGMs, see [20, 21].

On the whole, the advantage of these approaches over the previous process-based methods lies in the joint and concurrent appraisal of a variety of effects (each statistical model may consider an arbitrary number of variables to explain the shape of the observed network), with the drawback of reducing the contribution of each effect to a scalar quantity.

2.2 Reconstructing Structure

The second part of the double dichotomy (right-side in Table 1) relates to understanding the morphogenesis of the network itself. It may again be roughly divided into two broad categories, depending on whether approaches are based on a given growth process or on the topology of the network itself.

2.2.1 Using Processes

A myriad of models have been proposed to reconstruct network structure from normative assumptions. This is perhaps the most well-known and natural approach in statistical physics. At the core of these approaches lies generally a master equation or a master process featuring a certain number of key and oftentimes stylized ingredients. These ingredients correspond to an ideally small subset of canonical growth processes, defining the essential rules for adding—and, rarely, removing—nodes, links, and most importantly towards which types of nodes. The goal often consists in reproducing the observed connectivity (such as degree distributions), cohesiveness (such as clustering coefficients), or connectedness (such as component size distributions).

One of the earliest successful attempts at summarizing network morphogenesis with utterly simple processes consisted again in analytically solving simple PA based on node degree [15]. Models based on a general notion of PA have been extended in various directions: taking into account the age of nodes [33], their Euclidean distance [44, 101], their intrinsic fitness [25], their rank [38], or their activity [81]; formalizing a notion of competition between nodes to attract new links [17, 34, 36]; copying links from "prototype" nodes [60] or using random walks [94]; introducing preferences for transitive closure [54] or for specific groups of nodes (based on an a priori taxonomy [62] or an affiliation network [103]); or mixing structural PA with semantic PA (e.g., [70] who introduces the so-called degree-similarity model after observing that connected web pages are rather more similar, or [87] who mixes group-based PA and semantic PA). Group-based PA may also be found in models which describe the addition of groups rather than dyadic links, such as [46]: the network evolves through the iterative addition of teams and thus links between all their members, assuming a certain propensity to introduce newcomers and repeat past interactions.

Another class of models is based on link rewiring. One of the simplest versions was introduced by [98], who start with a ring lattice of fixed degree and reconnect links with a given probability p. This led to a discussion of the resulting structure in terms of low path length and high clustering coefficient, or "small-world." [29] later reproduced these two statistical features by adopting a distinct approach based on a rewiring process aimed at optimizing a global cost function, in a way inspired by [36].

Finally, a broad class of network models, especially in the social realm, falls into the category of *agent-based models* as soon as they rely on a relatively rich combination of processes. They generally aim at a specific application field which, in turn, requires detailed assumptions: as such, they typically offer a good combination of realism (they benefit from a stronger sociological grounding) and tractability (their study generally requires to resort to simulation). Examples of sophisticated models have been abundant in the social simulation literature from early on and are now present in a wide array of works at the interface with statistical physics and computational social science. It is way beyond the scope of this paper to attempt an overview of the wide diversity of agent-based network models. Let us nonetheless casually mention [40], who models the heterogeneous distribution of papers authored by scientists in a given field, and further reproduces the clustering of nodes in a semantic space, based on simple copying rules and the notion of quanta of knowledge called "kenes," by analogy with genes; [83], who build various social exchange network shapes by combining various agent decision heuristics and cognitive constraints; and [42] who reproduce blogger posting behavior and citation networks through a combination of random-walk-based generators and post-selection rules.

2.2.2 Using Structure

Reproducing graph structure directly from graph structure essentially means show-ing that some structural constraints entail the presence of other structural features—for instance, by demonstrating that a certain number of connected components or a strong proportion of some sort of triads follows from a given degree or subgraph distribution. Early attempts precisely focused on prescribing a power-law degree sequence [4] and, shortly thereafter, any degree sequence [77]. Several methods have later been proposed in the case of more sophisticated constraints, such as prescribed degree correlations [68], subgraph distributions [57], or recursive structures [63].

A typical challenge consists in being able to sample the space of graphs induced by a given set of constraints. Some approaches manage to provide a closed-form expression of several average statistical properties of the induced graph space, as has been done for the typical path length or average clustering coefficient by [78]. When this is not possible, an alternative consists in sampling the graph space through iterative exploration: the initial empirical graph is typically transformed by swapping pairs of edges while respecting the original constraint [41, 84]. This corresponds to a navigation in a meta-graph gathering all graphs of the target space. Beyond simple constraints, exhaustive navigation is usually impossible. [93] practically address this issue with an empirical sampling method denoted as "k-edge switching," iteratively swapping groups of k links in order to cover an increasingly large portion of a given graph space.

2.3 Combining Both: Evolutionary Models

In all four positions of the double dichotomy, the challenge generally consists in proposing one or several processes or constraints which will be key to explain network formation—be it transitivity, centrality, homophily, etc. The importance of such and such mechanism may be either assumed *a priori*, by looking at its effect on the network evolution, or verified *a posteriori*, by confirming its existence and appraising its shape during the network evolution. In all cases, intuition plays a key role. Yet, creating these models requires insights that may sometimes be unconventional.

To alleviate this dependence, evolutionary algorithms were recently used to automatically propose sets of mechanisms inferred from the observed structure. It differs from the above-mentioned methods in that it jointly uses the structure to reconstruct processes and the processes to reconstruct the structure. More precisely, network structure is used to devise link formation processes and, in turn and iteratively, these discovered processes are precisely used to reconstruct the structure.

Some of the earlier approaches introduced template models based on sets of possible specific actions (e.g., creating a link, rewire an edge, connecting to a random node, etc.). Actions have been organized in various manners: first as a fixed chart, resembling the typical structure of agent-based models [71], as a sequential list of variable size [13, 14, 49, 50] or, very recently, as a matrix whose weights describe the relative contribution of each action [10, 11]. In all these works, the evolutionary process aims at automatically (1) filling the template model with actions and (2) fitting the corresponding parameters. As is typical in evolutionary programming, it involves a fitness function which evaluates the resemblance between the empirical network and networks produced by the evolved model. Fitness functions rely on classical structural features (degree distributions, motifs, distance profiles, etc.). Models are iteratively evolved along increasing fitness values.

In parallel, we further proposed an original approach based on genetic programming and aimed at inferring arbitrarily complex combinations of elementary processes, construed as laws [72, 73].[1] We first introduced a generic vocabulary making it possible to describe network evolution in a unified framework, as an iterative process based on the likelihood of appearance of a link between two nodes, construed as a function on node properties in the currently evolving network (i.e., a form of generalized preferential attachment)—relying on structural features such as distance, connectivity, as well as non-structural characteristics. Representing these functions as trees enabled us to apply genetic programming techniques to evolve rules which are then used to generate network morphologies increasingly similar to the target, empirical network.

This technique may be denoted as "symbolic regression," for the goal is to use genetic programming to evolve free-form symbolic expressions rather than fitting parameters associated with fixed symbolic expressions: we *automatically* evolve realistic morphogenetic rules from a given instance of an empirical network, thereby symbolically regressing it. This strategy is inspired by the work of [90] who extract free-form scientific laws from experimental data. We first applied our method on kinship networks [72] which led to the publication of a much more general manuscript [73]. One remarkable result consists of the ability to *systematically* and *exactly* discover the laws of an Erdős–Rényi or Barabási–Albert generative process from a given stochastic instance. Distinct, realistic, and compact laws for a variety of social, physical, and biological networks could also be found.

We now describe in detail the core of the symbolic regression approach.

3   Symbolic Regression of Network Generators

We construe network generation as a stochastic process where edges are added iteratively, following some probability-based preference. Our approach is embedded

---

[1]In terms of potential applications, this approach has been evoked in the context of human connectome modeling [3, 18], as an alternative to conventional social simulation models [7] or to appraise matrimonial preferences from genealogical networks [74].

in a generalized preferential attachment framework centered around the notion of *generator* which is a scoring function providing a way to prefer some link over the others. A generator thus assigns a score s_{ij} to all edges (i, j). At each step of the network construction, a random sample S of candidate edges is drawn, among which a new edge is stochastically selected with a probability P_{ij} proportional to s_{ij} such that:

$$P_{ij} = \frac{s_{ij}}{\displaystyle\sum_{i', j' \in S} s_{i'j'}} \tag{1}$$

In practice, we forbid negative values and replace them with 0; in the special case where all weights are zero, they are all set to 1 for mathematical consistency.

In other words, generators implement an (arbitrarily complex) form of PA restricted to a random subset of links. Our core aim thus consists in designing a process able to automatically discover score computation functions s which yield networks comparable to a target empirical network. We construe generators as *tree-based computer programs* which represent mathematical expressions. Tree nodes are operators while leaves are variables and constants. Operators include classical arithmetic operations $\{+, -, *, /\}$, general-purpose mathematical functions: $\{x^y, e^x, \log, abs, \min, \max\}$, conditional expressions: $\{>, <, =, = 0\}$, and an affinity function (ψ, which we will further describe below). Variables are classical monadic or dyadic network measures which apply to the two nodes participating in the edge (i, j) to be scored: centrality degrees of each vertex (k_i and k_j), topological distance between the two vertices (d),[2] and their sequential identifiers (i and j, whose role we also discuss later on). We limit here the presentation of our approach to undirected networks with a fixed set of nodes, which fits our empirical material of Facebook ego-centered friendship networks. Nonetheless, it can straightforwardly be extended to directed networks (as in our original work) and the regular arrival of nodes.

This simple setting provides a language for describing generators and expressions which model and produce non-linear and non-centralized growth mechanisms. We now need a way to measure the similarity between the target network and generator-produced networks. This will provide the basis for defining the fitness function of our genetic approach. To this end, we first use a combination of distributions related to various topological aspects of the network, such as degree and PageRank [24] centralities, distance distributions, and triadic profiles [75]. We then compute dissimilarities between the respective distributions: for centralities, we apply the Earth mover's distance (EMD) [66], for the other distributions, we simply use ratio-based dissimilarity metrics. Of course, other metrics and dissimilarity measures

---

[2]To compute distances we use an heuristic based on a random walk, for (1) the exact computation is computationally intensive and, what is more, (2) new connections are also likely guided by a hop-by-hop navigation mechanism instead of an omniscient knowledge of the exact number of hops separating two given nodes.

may be used; we made these choices as a simple and intuitive trade-off between tractability and topological realism, which happens to work well.

By minimizing these dissimilarity measures, we get closer to the target network. This corresponds to a multi-objective optimization problem where some dissimilarities have to be minimized to the possible expense of others. We adopt a simple strategy by considering all dissimilarities in regard to the improvement over a random network. In other words, we divide the dissimilarity between the target network and a generated network by the dissimilarity between the target network and the average of 30 Erdős–Rényi (ER) random networks of the same size (same number of nodes and edges as the target). For a given metric, this means that if the dissimilarity between the target network and the ER average is, say, 5 and the distance from the target network to the generated network is 3, the ratio is 3/5. The smaller the ratio, the better the improvement—a ratio of 1 corresponds to no improvement. The evolutionary algorithm then aims at improving generators by minimizing the highest of these ratios. This defines our fitness function: the lower its value, the better the fitness.[3]

Our framework relies on a further feature: we allow node heterogeneity, i.e., we assume that not all nodes are (and thus behave) the same, irrespective of their structural position. Some actors of a social network may, for example, be intrinsically more likely to form ties with a specific class of actors. Here, we simply take heterogeneity into account through the sequential index of the node $i \in \{1, .., n\}$. These indices, considered as *identifiers*, may be used as a variable by a generator, and may thus introduce *a priori* distinctions in actor types. As we shall see, this element is key in the case of friendship networks where social circles play an essential structuring role.

Consider, for instance, the generator $s(i, j) = \frac{1}{i}$. It induces a probability of edge creation entirely determined by the identifier of one of its extremities. Nodes have distinct *a priori* propensities to originate connections, distributed following a hyperbolic curve. While integer identifiers may appear to introduce heterogeneity in very simplistic way, they can be combined with the other building blocks in an infinity of manners—and our results below show that indices were indeed used in sometimes creative ways.

Furthermore, index-based heterogeneity may be leveraged to define generators where certain vertices have natural affinity for each other. This brings us to the *affinity* function ψ, which uses the modulo operation to partition the identifier space into a certain number of groups. It relies on three operands: a constant, g, the number of groups, and two expressions, a and b, which are conditional outputs. If target and origin nodes i and j are equal modulo g, and thus belong to the same group (i.e., in case of "affinity"), the function returns a and b otherwise:

---

[3]ER is admittedly a basic null model. Yet, opting for a richer model is likely to induce bias: for instance, a fitness function based on a comparison with a configuration model would precisely incorporate target network degree distributions, making it impossible to directly approximate them.

$$\psi_g(i, j, a, b) = \begin{cases} a, & \text{if} (i \bmod g) \equiv (j \bmod g) \\ b, & \text{otherwise,} \end{cases} \qquad (2)$$

From now on, we consider i and j as implicit variables and denote the function as: $\psi_g(a, b)$.

Combining all these elements into an evolutionary loop makes it possible to generate plausible models for network generators, as summarized in Fig. 2.

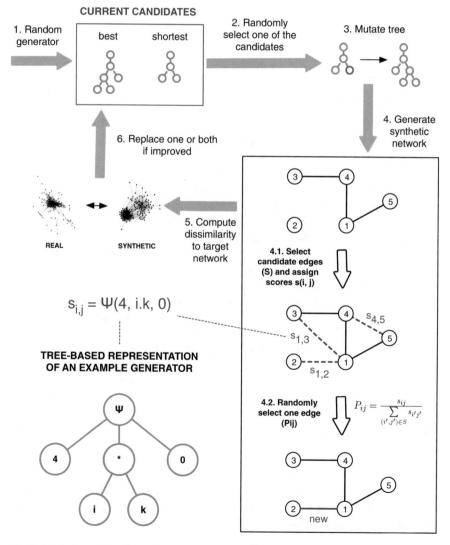

Fig. 2 Evolutionary loop including the synthetic network generation process. The outer part of this figure describes evolution at the generator population level, while the framed part on the right describes the evolution of a network for a given generator

Several runs with the same target network may generate different models—although they appear experimentally to converge onto similar behaviors. This leaves it to practitioners to select among the various options, conceivably by involving domain knowledge. A more objective consideration pertains to a trade-off between simplicity and precision. Since generators are essentially programs, complexity may be simply appraised through program length, an upper bound on the Kolmogorov complexity [76]. We thus apply a quantified version of Occam's Razor: all other things being equal, we also keep the model with the lowest program length that is within 10% of the best fitness.

4   Families of Network Generators

This approach provides the equivalent of an artificial scientist proposing plausible network models, replacing the intuition of the modeler. Using a biological analogy, it also makes it possible to discuss networks in terms of their plausible genotype (i.e., generator equations) rather than phenotypes (i.e., a series of topological traits).

Phenotypical traits assuredly provide the basis for appraising the quality of structural reconstruction and, by extension, for defining fitness functions attentive to such and such topological property [for an early yet already comprehensive review on the possible properties, see 31]. They also provide a good foundation for comparing networks with one another: a series of studies has indeed been devoted to defining network families by relying on triadic profiles [75], canonical analysis of various measures [31, section 19], adjacency matrix spectrum [35], blockmodeling [47], community structure [79], hierarchical structure [30], communication efficiency [43], and graphlets [100]. Note that this last method has been precisely used by [26] to phenotypically categorize the empirically networks we are dealing with here. Phenotypical traits have also been the target of evolutionary algorithms in [69], who symbolically regress formulas describing the phenotype of the network, e.g., finding an explicit expression for the diameter of various classes of networks as a function of the number of nodes, links, or some eigenvalues of the adjacency matrix.

By contrast, symbolic regression enables the comparison and categorization of networks based on their plausible underlying morphogenesis rules—as such a *genotypic categorization*. The core of the present contribution consists in applying our approach on a collection of networks of the same nature, unlike [73] which addresses a limited number of networks of different natures—biological, social, and man-made, both directed and undirected.

Here, we will exhibit families of generators, both in terms of their function and in terms of their expression. Their existence further suggests that a single mathematical expression and thus explanation may apply to a number of distinct empirical networks. In turn, it is thus even likely to correspond to a widespread class of actual generative behaviors.

4.1   Protocol

We use 238 anonymized ego-centered networks of Facebook friends which were randomly sampled from about 10,000 such networks collected in a large-scale online survey organized within a collaborative project called "Algopol" (consenting participants accepted to give access to their publication and network constitution history). Unlike other social networks such as Twitter, with concepts of "following" and "being followed," Facebook friend relationships are reciprocal and thus undirected. Furthermore, in ego networks, ego is by definition connected to every other node, so its presence would likely lead to more complex generators without any added explanatory power. We thus discard ego and all of their links.

For each network, we performed five evolutionary search runs. We then selected the generator discovered by the run that attained the highest fitness. This is a simple strategy to avoid low-quality local optima.

4.2   A Measure of Generator Dissimilarity

To identify families of generators and to visualize how similar they are in relation to each other, we start by introducing a measure of dissimilarity between pairs of generators. We understand the generator expression as the genotype and the network created using the generator as the phenotype. As in biology, different phenotypes can correspond to the same genotype. In our case, and beyond the intrinsic stochasticity of the generative process, this is trivially true because we can use the same generator to create networks of different sizes—both in numbers of nodes and edges. It is also true that different genotypes can create similar phenotypes. Notions of dissimilarity could be imagined on both the genotype and phenotype sides. On the genotype side, this could be a measure of program dissimilarity, for example, something akin to an edit distance. On the phenotype side, it could be a comparison of generated networks. We opted for the latter: we look for collections of generators that produce similar networks, and then check these groups to see if they contain regularities or competing explanations. In the end we propose a qualitative–quantitative analysis of families of generators.

Comparing networks is not a trivial task, and it becomes even harder for networks that do not have the same number of nodes and edges. With our generators, we are in a position to control this latter aspect. We use the generators to create synthetic networks that do not have the varied topologies of the ones they were derived from, but instead have a predetermined number of nodes and edges, facilitating subsequent comparison. We chose to generate networks of 1000 nodes and 10,000 edges, deemed to be large and dense enough for comparisons to be meaningful, and yet not so large that the task of comparing all pairs would become computationally intractable.

For the comparison itself, we employ a modified version of the fitness function that was used during the generator discovery process. The fitness function for undirected networks uses four distribution distance measures: k_d for degree; PR_d for PageRank; d_d for distance, and τ_d for the triadic profile. In that case, these measures are used to compare a synthetic network against the target network. Here, we will use them to compare pairs of synthetic networks created by the discovered generators. Being $n = 238$ the number of generators we consider four $n \times n$ matrices of pairwise distances, one for each measure: D_k, D_{PR}, D_d, and D_τ. To make these measures directly comparable, we produce normalized versions of each of these metrics in the following way:

$$D'_{i,j} = \frac{D_{i,j} - \min_{i'}(D_{i',j})}{\max_{i'}(D_{i',j}) - \min_{i'}(D_{i',j})} \tag{3}$$

The global dissimilarity function $\delta(i, j)$ is then simply the sum of the four normalized distances between two generated networks:

$$\delta(i, j) = D'_{k_{i,j}} + D'_{PR_{i,j}} + D'_{d_{i,j}} + D'_{\tau_{i,j}} \tag{4}$$

Notice that the above normalization process can lead to different estimations of the several distances depending on the direction because the normalization process is not symmetrical. We therefore finally use a symmetrized dissimilarity function σ' defined as $\sigma'(i, j) = \sigma(i, j) + \sigma(j, i)$.

4.3 Two-Dimensional Embedding and Families

To produce a visualization of the landscape of generators according to the above dissimilarity measure, we model these dissimilarities as distances in geometric space. We apply a metric *multi-dimensional scaling* [22] algorithm[4] (MDS) to map them into a two-dimensional space. Distances between pairs of points are set to match dissimilarity values as closely as possible.

We present the result of this two-dimensional embedding in Fig. 3.

We also performed a manual analysis, looking for patterns of similar generators in mathematical terms, i.e., at the level of the explicit formula. We identified 11 such strong patterns, and labeled every generator that conforms to one of them. We refer to sets of generators that conform to such patterns as *families* ($n' = 91$). The other ones are described as *unclassified* ($n'' = 147$).

---

[4]We used the metric MDS manifold embedding provided by the *scikit-learn* Python module.

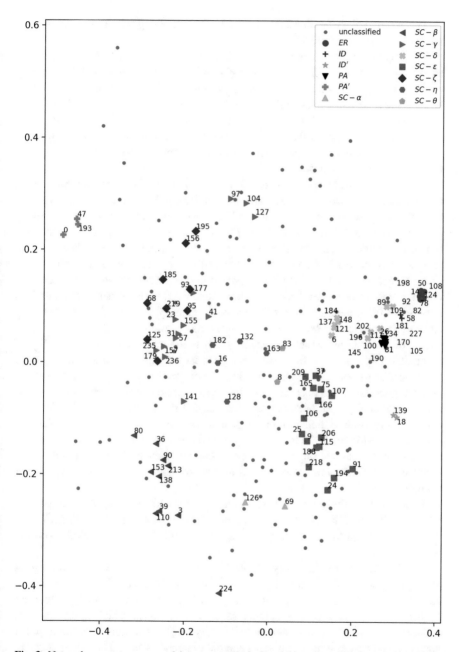

Fig. 3 Network generators mapped into a two-dimensional layout according to their pairwise distances. Different colors and shapes indicate families of generators that were manually identified as semantically similar. The legend shows the pattern that identifies each family

This manual classification is presented in Table 2, along with the actual genera-tors assigned to each family.[5]

The legend of Fig. 3 shows names that we gave to each family in Table 2, based on their main common primary mathematical features (we detail the meaning of these names below). A first interesting observation of this result is that the families are distributed in spatial clusters. Visual inspection makes it quite clear that mathematically close generators appear in similar regions of the 2D plane, some being much more spread than others. Another interesting point is that, although many generators are left unclassified, families are spread across most of the extent of the overall spatial distribution.

In the middle-right region of Fig. 3 we can find two families that correspond to well-known network models. The first is family **ER**, of the generators that are defined by some constant value c. They assign the same probability to every potential edge, and thus correspond to *Erdős–Rényi* random graphs. The second is family **PA**, of the generators that are defined by the degree variable k. They assign to each potential edge a probability that is proportional to the degree of either the origin or the target, and thus correspond to pure *preferential attachment* networks. It is interesting to observe that these two quintessential network formation explanations show up in our generator set, albeit in a small quantity. Further, they are relatively close to each other in respect to many other, more complex explanations. A third family of very simple generators (family **ID**) is the one where the probability of a potential edge is proportional to the sequential identifier of either the origin or the target. These generators are defined by the expression i. These are the simplest possible generators that take into account non-topological or exogenous features of nodes. This family is situated between the **ER** and **PA** families. Two other families exhibit expressions which roughly appear to be exponential versions of **ID** and **PA**. We named them **ID'** and **PA'**: they nonetheless behave very distinctly as the exponential induces a strong winner-takes-all effect on the highest value of the main variable (i or k). They are also situated in parts of the space distinct of their linear counterparts.

Notice that for these simple cases, although many of the generators are exactly the same, their positions do not coincide precisely in the spatial embedding. This is due to the fact that the generative process is stochastic, and some random variation is to be expected.

The first five families are very simple. The other eight have a very strong resem-blance with one another: they all use the affinity function, based on some constant number of affinity groups. This means that link dynamics is strongly influenced by the existence of a certain number of classes of nodes which likely matches underlying *social circles*; we denote this family as **SC**. A simple interpretation for this is indeed that ego networks are a sample of social groups that ego belongs to.

---

[5]Given the undirected nature of these networks, we simplify the notation for generators that use only variables from either the origin side or target side. Suppose we have the generator $3 \cdot k_i + d$; here, this is equivalent to $3 \cdot k_j + d$, so we simply write $3 \cdot k + d$.

Table 2 Generator expressions for each family

List of generator functions and corresponding network number — ⟨ID⟩

Family						
ER c	0.08 ⟨14⟩ $(\max(k_i, i) = 0 \rightarrow 0, 0.63)$ ⟨198⟩	0.88 ⟨50⟩	0.95 ⟨78⟩	54.6 ⟨82⟩	0.62 ⟨108⟩	6.0 ⟨124⟩
ID i	i ⟨58⟩	i ⟨109⟩				
ID' e^i	e^i ⟨18⟩	e^i ⟨139⟩				
PA k	k ⟨26⟩ k ⟨145⟩	k ⟨81⟩ k ⟨170⟩	k ⟨100⟩ k ⟨227⟩	k ⟨105⟩	k ⟨111⟩	k ⟨134⟩
PA' $k_j^{k_i}$ $k_i^{k_j}$	$k_j^{k_i}$ ⟨0⟩	$(\min(j, 0.66) > k_i \rightarrow j, e^{k_j})(\min((j=0, k_j, k_i), e^{k_j}))$ ⟨47⟩				$k_i^{k_j}$ ⟨193⟩
SC-α $\psi_g(k^s, c)$	$\psi_8(k_j^2, 0.62) - k_i$ ⟨69⟩		$\psi_7(k^3, 4)$ ⟨126⟩			
SC-β $\psi_g(e^k, > \tfrac{1}{2})$	$\psi_3(2^k, 0.48)$ ⟨3⟩ $\psi_4(e^k, 1)$ ⟨110⟩	$\psi_9(e^{k_i}, 0.49)$ ⟨36⟩ $\psi_8(e^k, d)$ ⟨138⟩	$\psi_4(e^k, 1.1)$ ⟨39⟩ $\psi_4(k, 0.67)^k$ ⟨153⟩	$\psi_5\left(\dfrac{e^{\max(k_i, k_j)}}{k_i}, k_i\right)$ ⟨80⟩ $\psi_5(e^k, 1.7)$ ⟨213⟩	$\psi_3(e^k, 2)$ ⟨224⟩	$\psi_5(e^k, 1)$ ⟨90⟩
SC-γ $\psi_g(k^B, \sim 0)$	$\psi_9(k^k, 0)$ ⟨23⟩ $\psi_3(e^k, 0)$ ⟨104⟩ $\psi_2(k_i \cdot e^{k_j}, 0)$ ⟨164⟩	$\psi_6(3^k, 0)$ ⟨31⟩ $\psi_3(2^k, 0)$ ⟨127⟩ $\psi_4(e^k, 0)$ ⟨177⟩	$\psi_4(4 \cdot k^5, 0)$ ⟨41⟩ $\psi_6(e^{\psi_5(1,k)}, 0) + 0.07$ ⟨141⟩ $\psi_5(k^7, 0.01)$ ⟨235⟩	$\psi_8(k^k, 0)$ ⟨57⟩ $\psi_5(e^k, 0.03)$ ⟨236⟩	$\psi_3(e^{k_i+k_j}, 0.05)$ ⟨97⟩ $\psi_7(e^k, 0)$ ⟨155⟩	$\psi_4(e^k, 0.06)$ ⟨157⟩

(continued)

Table 2 (continued)

Family	List of generator functions and corresponding network number					⟨ID⟩
SC-δ $\psi_g(e^i, *)$	$\psi_4(e^i, e^{k_j})$ ⟨6⟩	$\psi_4(i^j, k_j)$ ⟨89⟩	$\psi_2(j^i, k_i)$ ⟨92⟩	$\psi_3(e^i, k_i)$ ⟨121⟩	$\psi_3(e^i, e^7)$ ⟨137⟩	$\psi_3(e^i, 1)$ ⟨148⟩
	$\psi_2(9^i, 9^9)$ ⟨181⟩	$\psi_3(e^i, j)$ ⟨184⟩	$\psi_3(e^{i+j-d}, e^5)$ ⟨196⟩		$\psi_4(9^i, 9)$ ⟨202⟩	
SC-ε $\psi_g(ik, *)$	$9\psi_3(ik_i, 2k_i)$ ⟨9⟩	$\psi_4(ik_j, 6k_j)$ ⟨24⟩	$\psi_5(ik_i, k_i)$ ⟨25⟩	$\psi_9(ik_i, 0.1k_i)$ ⟨37⟩	$\psi_2(ik_i, k_i)$ ⟨75⟩	$\psi_7(ik_i, 7k_i)$ ⟨91⟩
	$\psi_6(ik_i, 0.44k_i)$ ⟨106⟩	$\psi_4(jk_i, 0.38)$ ⟨107⟩	$\psi_3(jk_j, k_j)$ ⟨115⟩	$\psi_4(i \log(k_i), 0)$ ⟨165⟩		$\psi_3(jk_i, \frac{k_i}{4})$ ⟨166⟩
	$\left(\frac{k_j k_i}{0.66} + d\right)\psi_4(j, 0.61)$ ⟨188⟩		$\psi_3(ik_j, 2k_j)$ ⟨194⟩	$\psi_3(ik_j, k_j)$ ⟨206⟩	$\psi_3(ik_i, 0)$ ⟨209⟩	$\psi_4(ik_i, 3k_i)$ ⟨218⟩
SC-ζ $\psi_g(i^k, *)$	$\psi_7(i, 0)^{k_j}$ ⟨68⟩	$\frac{7}{d}\psi_4(i^{k_i}, 0.48)$ ⟨93⟩	$\psi_4\left(\frac{k_j}{k_j}, 0.18\right)$ ⟨95⟩	$\psi_8(i^{k_i}, 2)$ ⟨125⟩	$\psi_4(i^{k_i}, 0)$ ⟨156⟩	$\psi_4(\frac{1}{6}i^{k_i}, d)$ ⟨179⟩
	$\psi_9(dj^{k_i}, 0)$ ⟨185⟩	$\psi_{min(i,4)}(i^{k_i}, 0)$ ⟨195⟩		$\psi_5(9j^{k_i}, 0.03)$ ⟨219⟩		
SC-η $\psi_g(ik^2, *)$	$\psi_5((ik_i)^2, i)$ ⟨16⟩	$\psi_5(ik_i^2, 6)$ ⟨128⟩	$\psi_4(2980.96k^2, 2k)$ ⟨132⟩		$\psi_2(ik_j^2, k_j^2)$ ⟨163⟩	
	$\psi_7(\psi_i(0.5, k_j^2), 0)$ ⟨182⟩					
SC-θ $\psi_g(k, 0) - 1$	$\psi_4(k, 0) - 0.99$ ⟨8⟩		$\psi_7(k, 0) - 0.93$ ⟨83⟩			

c represents a constant value, s a small exponent, B a big exponent, and $*$ is used as a placeholder for an arbitrary expression

For example: school friends, family, work colleagues, and so on. It makes sense that these groups are much more densely connected within themselves than between them, as they correspond to separate social spheres. The affinity function provides a very straightforward way of generating this type of linking behavior. The constant number of groups present in the first parameter of affinity functions represents an estimation of the number of social groups that ego belongs to. In our previous work [73] we included one Facebook ego network in the diverse set of networks used, and the generator found for it was also based on an affinity function. In fact, under the typology we present here, it would be classified as an **SC**-ϵ generator. From the biological, social, and technological networks analyzed in that work, the Facebook ego network was the only one based on an affinity function with a constant number of groups. This presents us with additional empirical evidence that this is in fact a characteristic signature of ego-centered social networks.

To illustrate further these families, we provide a few visual examples of network generators in Table 3. For each selected generator of a given family, we put along the original empirical network and its reconstruction using the same number of nodes. Spatialization follows a force-directed layout. The number of social circles parameterized on ψ may be seen to be faithful to the original number of clusters in the real network.

SC families differ in the linking behavior for nodes deemed to belong to the same group. Some of them are purely based on topological factors (families α, β, γ, and θ), one only on exogenous factors (family δ), and some on a combination of both (families ϵ, ζ, and η).

Table 3 Visual representation of some empirical ego networks (top row) with their reconstruction (bottom row), for a selection of evoked families

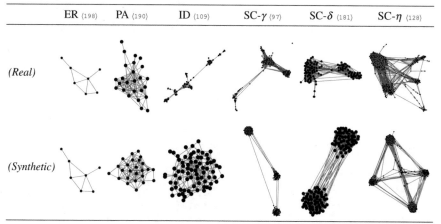

	ER (198)	PA (190)	ID (109)	SC-γ (97)	SC-δ (181)	SC-η (128)
(Real)						
(Synthetic)						

ER, PA, and ID are featured; each of the three main subfamilies of SC are also present (generators 97, 181, and 128 are all based on an affinity function of parameters 3, 2, and 5, respectively). Note that three of the empirical networks (109, 128, 181) feature very small disconnected components, gathering no more than a handful of nodes which have not been drawn here for clarity purposes

The largest family is ϵ, which assigns probability of in-group links as a linear combination of current degree (k) and exogenous factors (i). The second largest family by number of generators found is family γ, and it is also the one that is the most spread in the spatial embedding. In this family, the probability of in-group connections is purely driven by topology, as an exponential of the current degree of one of the nodes. We can think of it as a form of super-preferential attachment within social circles—current popularity within the group is highly rewarded. For most cases, the probability of connection between groups is given by a relatively small constant, and for a few it is zero.

Some questions remain. Why are some generators so simple, and why are more than half of the generators so diverse that they cannot be classified into families? In an attempt to attain a better understanding, we created boxplots of the distributions of node and edge counts for the underlying networks per family, as well as for all generators, and for classified and unclassified generators. These plots are presented in Fig. 4, as well as a stacked plot of family ratio per percentile of network density. Some interesting facts are revealed.

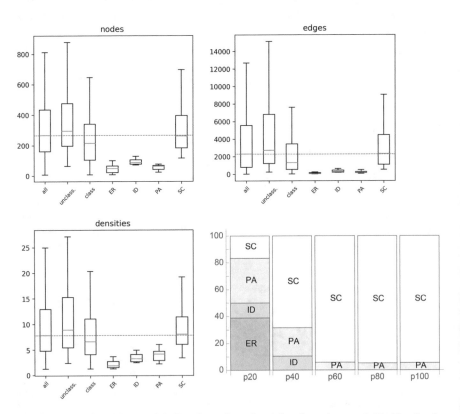

Fig. 4 *Top panel and bottom-left:* Boxplots of numbers of nodes, edges, and densities for the underlying networks of the various families, as well as all, unclassified and classified. Horizontal dashed line indicates overall median. *Bottom-right:* Stacked plot of family ratio per percentile of network density

The families of simpler generators (**ER, ID, ID′, PA,** and **PA′**) have both node and edge counts well below the median. This could indicate that these simple generators correspond to cases where there is not enough data to form a more complex theory. The simplest underlying behavior is captured, corresponding precisely to the simple archetypal explanations of preferential attachment and random behavior. Maybe these networks are small because the corresponding user is not very active, or does not have many social connections, or maybe because they joined recently and the networks are at their initial stages of growth. When the latter case is true, our results seem to indicate that they may be assignable to a more complex family when they develop more. Under this assumption, families **SC** paint here the more relevant part of the picture of network growth dynamics.

The unclassified set corresponds to networks that are slightly larger than the mean, both in numbers of nodes and edges. From this observation we formulate two hypotheses. The first one is that the unclassified set really does correspond to a complex variety of behaviors. It could be that, given one or two orders of magnitude more ego networks, more families would be found. The second one is that it is more difficult for evolutionary search to find simple generators for these larger networks, but that given more runs, they would emerge.

In an attempt to test the second hypothesis, in Fig. 5 we plot the best fitnesses achieved for the underlying networks, again per family, all generators, classified and unclassified. Here we find that generators of the **SC** family attain slightly better fitness (both for the median and worst cases) than generators of unclassified networks. This lends some credence to the second hypothesis. Furthermore, we

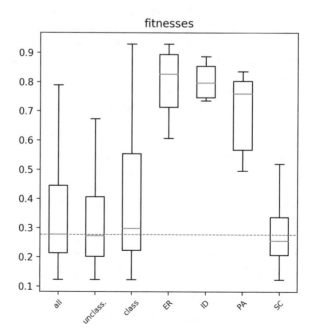

Fig. 5 Boxplots of best fitnesses achieved (lower is better) for the underlying networks of the various families, as well as all, unclassified and classified. Horizontal dashed line indicates overall median

observed that for the entire **SC** family, the generator with a simple pattern was only found once, and it always had the best fitness of the five runs. It seems thus likely that, given more evolutionary search runs per generator, at least part of the unclassified networks would fall into a family.

It is not possible to know if the families are exhaustive or the simplest that could be found. Investing more computational power on this problem could always yield simpler yet harder to find explanations, both for the classified and unclassified cases. It could also show unclassified networks to belong to a known family, or to a new family. As with many heuristic methods, the best we can do is trust some stability criteria (e.g., stop at a certain number of runs without anything new being found).

5 Final Remarks

We believe that several interesting explorations can stem from the symbolic regression of network generators. After the research work presented in this chapter, we are left encouraged by the potential of a genotype-based approach in describing families of generators. To move to a larger scale, it is necessary to go further in the methods to identify similar generators at the semantic, i.e., mathematical level. This is a hard but exciting computer science problem.

It would also be interesting to map the space of possible generators by searching not for generators that target specific networks, but instead that attempt to generate networks as divergent as possible from those already known. Combining this exploration with family identification could lead to insights related to the families of generators found in different scientific fields and types of phenomena, as well as to families that do not correspond to networks found in any empirical data. This could reveal potentially interesting network designs, as was the case with the evolved radio antennas.

Our current method assumes homogeneous behavior across the network. Hybrid methods combining community detection with symbolic regression could lead, in certain cases, to more powerful explanations with different generator expressions per sub-network.

Another important challenge is that of targeting dynamic networks. This will require a fitness function that takes into account different stages of growth of a target network, and that leads to generators that can be validated to produce not only a plausible state of the network at a certain stage, but a plausible growth process overall.

Acknowledgements We are grateful to the members of our "Algopol" project team (ANR-12-CORD-0018) who organized most of the Facebook survey, including Irène Bastard, Dominique Cardon, Raphaël Charbey, Guilhem Fouetillou, Christophe Prieur, and Stéphane Raux. We further acknowledge interesting discussions with Jean-Philippe Cointet and David Fourquet regarding generative families, as well as the constructive feedback of our anonymous reviewers. This paper has been partially supported by the "Algodiv" grant (ANR-15-CE38-0001) funded by the ANR (French National Agency of Research).

References

1. Acar, E., Dunlavy, D.M., Kolda, T.G.: Link prediction on evolving data using matrix and tensor factorizations. In: Proceedings of ICDMW'09, IEEE International Conference on Data Mining Workshops, pp. 262–269. IEEE, Piscataway (2009)
2. Adar, E., Zhang, L., Adamic, L.A., Lukose, R.M.: Implicit structure and the dynamics of blogspace. In: Workshop on the Weblogging Ecosystem, 13th International World Wide Web Conference (2004)
3. Adolphs, R.: The unsolved problems of neuroscience. Trends Cogn. Sci. 19(4), 173–175 (2015)
4. Aiello, W., Chung, F., Lu, L.: A random graph model for massive graphs. In: Proceedings ACM 32nd Annual ACM Symposium on Theory of Computing, pp. 171–180. ACM, New York (2000)
5. Al Hasan, M., Chaoji, V., Salem, S., Zaki, M.: Link prediction using supervised learning. In: SDM: Workshop on Link Analysis, Counter-terrorism and Security (2006)
6. Al Hasan, M., Zaki, M.J.: A survey of link prediction in social networks. In: Aggarwal, C.C. (ed) Social Network Data Analytics, pp. 243–275. Springer, Boston (2011)
7. Amblard, F., Bouadjio-Boulic, A., Gutiérrez, C.S., Gaudou, B.: Which models are used in social simulation to generate social networks? A review of 17 years of publications in JASSS. In: Winter Simulation Conference (WSC), 2015, pp 4021–4032. IEEE, Piscataway (2015)
8. Anderson, C.J., Wasserman, S., Faust, K.: Building stochastic blockmodels. Soc. Net. 14, 137–161 (1992)
9. Anderson, C.J., Wasserman, S., Crouch, B.: A p* primer: logit models for social networks. Soc. Net. 21, 37–66 (1999)
10. Arora, V., Ventresca, M.: A multi-objective optimization approach for generating complex networks. In: Companion Proceedings of GECCO'16 18th Genetic and Evolutionary Computation Conference, pp. 15–16. ACM, New York (2016)
11. Arora, V., Ventresca, M.: Action-based modeling of complex networks. Sci. Rep. 7, 6673 (2017)
12. Avena-Koenigsberger, A., Goñi, J., Solé, R., Sporns, O.: Network morphospace. J. R. Soc. Interface 12(103), 20140881 (2015)
13. Bailey, A., Ventresca, M., Ombuki-Berman, B.: Automatic generation of graph models for complex networks by genetic programming. In: Proceedings GECCO'12 14th ACM Annual Conference on Genetic and Evolutionary Computation, pp. 711–718. ACM, New York (2012)
14. Bailey, A., Ventresca, M., Ombuki-Berman, B.: Genetic programming for the automatic inference of graph models for complex networks. IEEE Trans. Evol. Comput. 18(3), 405–419 (2014)
15. Barabási, A.L., Albert, R.: Emergence of scaling in random networks. Science 286, 509–512 (1999)
16. Barabási, A.L., Jeong, H., Ravasz, E., Neda, Z., Vicsek, T., Schubert, A.: Evolution of the social network of scientific collaborations. Physica A 311, 590–614 (2002)
17. Berger, N., Borgs, C., Chayes, J.T., D'Souza, R.M., Kleinberg, R.D.: Competition-induced preferential attachment. In: Proceedings of the 31st International Colloquium on Automata, Languages and Programming, pp. 208–221 (2004)
18. Betzel, R.F., Avena-Koenigsberger, A., Goñi, J., He, Y., De Reus, M.A., Griffa, A., Vértes, P.E., Mišic, B., Thiran, J.P., Hagmann, P., et al.: Generative models of the human connectome. Neuroimage 124, 1054–1064 (2016)
19. Bliss, C.A., Frank, M.R., Danforth, C.M., Dodds, P.S.: An evolutionary algorithm approach to link prediction in dynamic social networks. J. Comput. Sci. 5(5), 750–764 (2014)
20. Block, P., Stadtfeld, C., Snijders, T.A.B.: Forms of dependence: comparing SAOMs and ERGMs from basic principles. Sociol. Methods Res. 48(1) (2016)
21. Block, P., Koskinen, J., Hollway, J., Steglich, C., Stadtfeld, C.: Change we can believe in: comparing longitudinal network models on consistency, interpretability and predictive power. Soc. Net. 52, 180–191 (2018)

22. Borg, I., Groenen, P.J.F.: Modern Multidimensional Scaling: Theory and Applications. Springer Science & Business Media, New York (2005)
23. Brennecke, J., Rank, O.N.: The interplay between formal project memberships and informal advice seeking in knowledge-intensive firms: a multilevel network approach. Soc. Net. **44**, 307–318 (2016)
24. Brin, S., Page, L.: The anatomy of a large-scale hypertextual web search engine. Comput. Netw. ISDN Syst. **30**(1–7), 107–117 (1998)
25. Caldarelli, G., Capocci, A., De Los Rios, P., Munoz, M.A.: Scale-free networks from varying vertex intrinsic fitness. Phys. Rev. Lett. **89**(25), 258702 (2002)
26. Charbey, R., Prieur, C.: Graphlet-based characterization of many ego networks. hal-01764253v2 (2018)
27. Clauset, A., Moore, C., Newman, M.E.J.: Hierarchical structure and the prediction of missing links in networks. Nature **453**, 98–101 (2008)
28. Cointet, J.P., Roth, C.: Local networks, local topics: structural and semantic proximity in blogspace. In: Proceedings 4th ICWSM AAAI International Conference on Weblogs and Social Media, pp. 223–226. AAAI, Menlo Park (2010)
29. Colizza, V., Banavar, J.R., Maritan, A., Rinaldo, A.: Network structures from selection principles. Phys. Rev. Lett. **92**(19), 198701 (2004)
30. Corominas-Murtra, B., Goñi, J., Solé, R.V., Rodríguez-Caso, C.: On the origins of hierarchy in complex networks. PNAS **110**(33), 13316–13321 (2013)
31. da Fontoura Costa, L., Rodrigues, F.A., Travieso, G., Villas Boas, P.R.: Characterization of complex networks: a survey of measurements. Adv. Phys. **56**(1), 167–242 (2007)
32. de Solla Price, D.J.: A general theory of bibliometric and other cumulative advantage processes. J. Am. Soc. Inf. Sci. **27**(5–6), 292–306 (1976)
33. Dorogovtsev, S.N., Mendes, J.F.F.: Evolution of networks with aging of sites. Phys. Rev. E **62**, 1842–1845 (2000)
34. D'Souza, R.M., Borgs, C., Chayes, J.T., Berger, N., Kleinberg, J.T.: Emergence of tempered preferential attachment from optimization. PNAS **104**, 6112–6117 (2007)
35. Estrada, E.: Topological structural classes of complex networks. Phys. Rev. E **75**(1), 016103 (2007)
36. Fabrikant, A., Koutsoupias, E., Papadimitriou, C.H.: Heuristically optimized trade-offs: a new paradigm for power laws in the internet. In: ICALP '02: Proceedings of the 29th International Colloquium on Automata, Languages and Programming, London, UK, pp. 110–122, Springer, Berlin (2002). ISBN 3-540-43864-5
37. Fienberg, S.E., Meyer, M.M., Wasserman, S.S.: Statistical analysis of multiple sociometric relations. J. Am. Stat. Assoc. **80**(389), 51–67 (1985)
38. Fortunato, S., Flammini, A., Menczer, F.: Scale-free network growth by ranking. Phys. Rev. Lett. **96**, 218701 (2006)
39. Frank, O., Strauss, D.: Markov graphs. J. Am. Stat. Assoc. **81**(395), 832–842 (1986)
40. Gilbert, N.: A simulation of the structure of academic science. Sociol. Res. Online **2**(2), 1–15 (1997)
41. Gkantsidis, C., Mihail, M., Zegura, E.W.: The markov chain simulation method for generating connected power law random graphs. In: Proceedings 5th Workshop on Algorithm Engineering and Experiments (ALENEX) (2003)
42. Goetz, M., Leskovec, J., McGlohon, M., Faloutsos, C.: Modeling blog dynamics. In: ICWSM 2009 Proceedings 3rd International AAAI Conference on Weblogs and Social Media, AAAI, Menlo Park (2009)
43. Goñi, J., Avena-Koenigsberger, A., de Mendizabal, N.V., van den Heuvel, M.P., Betzel, R.F., Sporns, O.: Exploring the morphospace of communication efficiency in complex networks. PLoS One **8**(3), e58070 (2013)
44. Guimerà, R., Amaral, L.A.N.: Modeling the world-wide airport network. Eur. Phys. J. B **38**, 381–385 (2004)
45. Guimerà, R., Sales-Pardo, M.: Missing and spurious interactions and the reconstruction of complex networks. PNAS **106**(52), 22073–22078 (2009)

46. Guimera, R., Uzzi, B., Spiro, J., Nunes Amaral, L.A.: Team assembly mechanisms determine collaboration network structure and team performance. Science **308**, 697–702 (2005)
47. Guimerà, R., Sales-Pardo, M., Amaral, L.A.N.: Classes of complex networks defined by role-to-role connectivity profiles. Nat. Phys. **3**, 63–69 (2007)
48. Hanneke, S., Fu, W., Xing, E.P.: Discrete temporal models of social networks. Elect. J. Stat. **4**, 585–605 (2010)
49. Harrison, K.R., Ventresca, M., Ombuki-Berman, B.M.: Investigating fitness measures for the automatic construction of graph models. In: Mora, A., Squillero, G. (eds) EvoApplications 2015 Applications of Evolutionary Computation. LNCS, vol. 9028, pp. 189–200. Springer, Berlin (2015)
50. Harrison, K.R., Ventresca, M., Ombuki-Berman, B.M.: A meta-analysis of centrality measures for comparing and generating complex network models. J. Comput. Sci. **17**, 205–215 (2016)
51. Holland, P., Leinhardt, S.: A dynamic model for social networks. J. Math. Soc. **5**, 5–20 (1977)
52. Holland, P.W., Leinhardt, S.: An exponential family of probability distributions for directed graphs. J. Am. Stat. Assoc. **76**(373), 33–65 (1981)
53. Holland, P.W., Laskey, K.B., Leinhardt, S.: Stochastic blockmodels: first steps. Soc. Net. **5**, 109–137 (1983)
54. Holme, P., Kim, B.J.: Growing scale-free networks with tunable clustering. Phys. Rev. E **65**, 026107 (2002)
55. Hornby, G., Globus, A., Linden, D., Lohn, J.: Automated antenna design with evolutionary algorithms. In: Space 2006, AIAA SPACE Forum, pp. 1–8 (2006)
56. Jeong, H., Néda, Z., Barabási, A.L.: Measuring preferential attachment for evolving networks. Europhys. Lett. **61**(4), 567–572 (2003)
57. Karrer, B., Newman, M.E.J.: Random graphs containing arbitrary distributions of subgraphs. Phys. Rev. E **82**, 066118 (2010)
58. Koskinen, J.H., Snijders, T.A.B.: Bayesian inference for dynamic social network data. J. Statist. Plann. Inference **137**(12), 3930–3938 (2007)
59. Kossinets, G., Watts, D.J.: Empirical analysis of an evolving social network. Science **311**, 88–90 (2006)
60. Kumar, R., Raghavan, P., Rajagopalan, S., Sivakumar, D., Tomkins, A., Upfal, E.: Stochastic models for the web graph. In: IEEE 41st Annual Symposium on Foundations of Computer Science (FOCS), p. 57. IEEE Computer Society, Washington (2000)
61. Leskovec, J., Horvitz, E.: Planetary-scale views on a large instant-messaging network. In: Proceedings WWW'08 17th International Conference World Wide Web, pp. 915–924. ACM, New York (2008)
62. Leskovec, J., Kleinberg, J., Faloutsos, C.: Graphs over time: densification laws, shrinking diameters and possible explanations. In: Proceedings of the 11th ACM SIGKDD International Conference on Knowledge Discovery and Data Mining, pp. 177–187. ACM, New York (2005)
63. Leskovec, J., Chakrabarti, D., Kleinberg, J., Faloutsos, C., Ghahramani, Z.: Kronecker graphs: an approach to modeling networks. J. Mach. Learn. Res. **11**, 985–1042 (2010)
64. Lewis, K., Gonzalez, M., Kaufman, J.: Social selection and peer influence in an online social network. PNAS **109**(1), 68–72 (2012)
65. Liben-Nowell, D., Kleinberg, J.: The link prediction problem for social networks. In: CIKM '03: Proceedings of the 12th International Conference on Information and Knowledge Management, pp. 556–559. ACM Press, New York (2003)
66. Ling, H., Okada, K.: An efficient earth mover's distance algorithm for robust histogram comparison. IEEE Trans. Pattern Anal. Mach. Intell. **29**(5), 840–853 (2007)
67. Lü, L., Zhou, T.: Link prediction in complex networks: a survey. Physica A **390**, 1150–1170 (2011)
68. Mahadevan, P., Krioukov, D., Fall, K., Vahdat, A.: Systematic topology analysis and generation using degree correlations. In: Proceedings SIGCOMM'06 ACM International Conference on Applications, Technologies, Architectures, and Protocols for Computer Communications, pp. 135–146. ACM, New York (2006)

69. Märtens, M., Kuipers, F., Mieghem, P.V.: Symbolic regression on network properties. In: McDermott, J., Castelli, M., Sekanina, L., Haasdijk, E., García-Sánchez, P., (edts.) Proceedings EuroGP 2017 Genetic Programming. LNCS, vol. 10196. Springer, Berlin (2017)
70. Menczer, F.: Evolution of document networks. PNAS **101**(S1), 5261–5265 (2004)
71. Menezes, T.: Evolutionary modeling of a blog network. In: Proceedings CEC'2011 IEEE Congress on Evolutionary Computation, pp. 909–916. IEEE, Piscataway (2011)
72. Menezes, T., Roth, C.: Automatic discovery of agent-based models: an application to social anthropology. Advances in Complex Systems **16**(7), 1350027 (2013)
73. Menezes, T., Roth, C.: Symbolic regression of generative network models. Sci. Rep. **4**, 6284 (2014)
74. Menezes, T., Gargiulo, F., Roth, C., Hamberger, K.: New simulation techniques in kinship network analysis. Struct. Dynam. e-J. Anthropol. Related Sci. **9**(2), 180–209 (2016)
75. Milo, R., Itzkovitz, S., Kashtan, N., Levitt, R., Shen-Orr, S., Ayzenshtat, I., Sheffer, M., Alon, U.: Superfamilies of evolved and designed networks. Science **303**(5663), 1538–1542 (2004)
76. Ming, L., Vitányi, P.: An Introduction to Kolmogorov Complexity and Its Applications. Springer, Heidelberg (1997)
77. Newman, M.E.J., Strogatz, S., Watts, D.: Random graphs with arbitrary degree distributions and their applications. Phys. Rev. E **64**(026118) (2001)
78. Newman, M.E.J., Strogatz, S.H., Watts, D.J.: Random graphs models of social networks. PNAS **99**, 2566–2572 (2002)
79. Onnela, J.P., Fenn, D.J., Reid, S., Porter, M.A., Mucha, P.J., Fricker, M.D., Jones, N.S.: Taxonomies of networks from community structure. Phys. Rev. E **86**, 036104 (2012)
80. Papadopoulos, F., Kitsak, M., Serrano, M.Á., Boguná, M., Krioukov, D.: Popularity versus similarity in growing networks. Nature **489**, 537–540 (2012)
81. Perra, N., Gonçalves, B., Pastor-Satorras, R., Vespignani, A.: Activity driven modeling of time varying networks. Sci. Rep. **2**(469) (2012)
82. Powell, W.W., White, D.R., Koput, K.W., Owen-Smith, J.: Network dynamics and field evolution: the growth of interorganizational collaboration in the life sciences. Am. J. Sociol. **110**(4), 1132–1205 (2005)
83. Pujol, J.M., Flache, A., Delgado, J., Sangüesa, R.: How can social networks ever become complex? Modelling the emergence of complex networks from local social exchanges. J. Art. Soc. Soc. Sim. **8**(4) (2005)
84. Rao, A., Jana, R., Bandyopadhyay, S.: A markov chain Monte Carlo method for generating random (0,1)-matrices with given marginals. Sankhya: The Indian J. Stat. A **58**, 225–242 (1996)
85. Robins, G., Pattison, P., Kalish, Y., Lusher, D.: An introduction to Exponential Random Graph (p*) Models for social networks. Soc. Net. **29**(2), 173–191 (2007)
86. Roth, C.: Generalized preferential attachment: towards realistic socio-semantic network models. In: ISWC 4th International Semantic Web Conference, Workshop on Semantic Network Analysis, CEUR-WS Series (ISSN 1613-0073), vol. 171, pp. 29–42. Galway, Ireland (2005)
87. Roth, C.: Co-evolution in epistemic networks – reconstructing social complex systems. Struct. Dynam. e-J. Anthropol. Related Sci. **1**(3). Article 2 (2006)
88. Rowe, M., Stankovic, M., Alani, H.: Who will follow whom? Exploiting semantics for link prediction in attention-information networks. In: Cudré-Mauroux, P., Heflin, J., Sirin, E., Tudorache, T., Euzenat, J., Hauswirth, M., Parreira, J.X., Hendler, J., Schreiber, G., Bernstein, A., Blomqvist, E. (eds) Proceedings ISWC'12 11th International Semantic Web Conference Part I. LNCS, vol. 7649, pp. 476–491. Springer, Berlin (2012)
89. Sarkar, P., Chakrabarti, D., Jordan, M.: Nonparametric link prediction in large scale dynamic networks. Electron. J. Stat. **8**(2), 2022–2065 (2014)
90. Schmidt, M., Lipson, H.: Distilling free-form natural laws from experimental data. Science **324**(5923), 81–85 (2009)
91. Snijders, T.A.B.: The statistical evaluation of social networks dynamics. Sociol. Methodol. **31**, 361–395 (2001)

92. Snijders, T.A.B., Steglich, C., Schweinberger, M.: Modeling the co-evolution of networks and behavior. In: van Montfort, K., Oud, H., Satorra, A. (eds) Longitudinal Models in the Behavioral and Related Sciences, pp. 41–71. Lawrence Erlbaum, Mahwah (2007)
93. Tabourier, L., Roth, C., Cointet, J.P.: Generating constrained random graphs using multiple edge switches. ACM J. Exp. Algorithmics 16(1.7) (2011)
94. Vázquez, A.: Growing network with local rules: preferential attachment, clustering hierarchy, and degree correlations. Phys. Rev. E 67, 056104 (2003)
95. Wang, P., Robins, G., Pattison, P., Lazega, E.: Exponential Random Graph Models for multilevel networks. Soc. Net. 35, 96–115 (2013)
96. Wasserman, S.: Analyzing social networks as stochastic processes. J. Am. Stat. Assoc. 75(370), 280–294 (1980)
97. Wasserman, S., Pattison, P.: Logit models and logistic regressions for social networks: I. an introduction to markov graphs and p*. Psychometrika 61(3), 401–425 (1996)
98. Watts, D.J., Strogatz, S.H.: Collective dynamics of 'small-world' networks. Nature 393, 440–442 (1998)
99. Yang, Y., Lichtenwalter, R.N., Chawla, N.V.: Evaluating link prediction methods. Knowl. Inf. Syst. 45(3), 751–782 (2015)
100. Yaveroglu, Ö.N., Malod-Dognin, N., Devis, D., Levnajic, Z., Janjic, V., Karapandza, R., Stojmirovic, A., Przulj, N.: Revealing the hidden language of complex networks. Sci. Rep. 4(4547) (2014)
101. Yook, S.H., Jeong, H., Barabási, A.L.: Modeling the internet's large-scale topology. PNAS 99(21), 13382–13386 (2002)
102. Yuan, G., Murukannaiah, P.K., Zhang, Z., Singh, M.P.: Exploiting sentiment homophily for link prediction. In: Proceedings RecSys '14 8th ACM Conference on Recommender Systems, pp. 17–24. ACM, New York (2014)
103. Zheleva, E., Sharara, H., Getoor, L.: Co-evolution of social and affiliation networks. In: Proceedings ACM SIGKDD'09 15th International Conference on Knowledge Discovery and Data Mining, pp. 1007–1015. ACM, New York (2009)

Modeling User Dynamics in Collaboration Websites

Patrick Kasper, Philipp Koncar, Simon Walk, Tiago Santos,
Matthias Wölbitsch, Markus Strohmaier, and Denis Helic

Abstract Numerous collaboration websites struggle to achieve self-sustainability—a level of user activity preventing a transition to a non-active state. We know only a little about the factors which separate sustainable and successful collaboration websites from those that are inactive or have a declining activity. We argue that modeling and understanding various aspects of the evolution of user activity in such systems is of crucial importance for our ability to predict and support success of collaboration websites. Modeling user activity is not a trivial task to accomplish due to the inherent complexity of user dynamics in such systems. In this chapter, we present several approaches that we applied to deepen our understanding of user dynamics in collaborative websites. Inevitably, our approaches are quite heterogeneous and range from simple time-series analysis, towards the application of dynamical systems, and generative probabilistic methods. Following some of our initial results, we argue that the selection of methods to study user dynamics strongly depends on the type of collaboration systems under investigation as well as on the research questions that we ask about those systems. More specifically, in this chapter we show our results of (1) the analysis of nonlinearity of user activity time-series, (2) the application of classical dynamical systems to model user motivation and peer influence, (3) a range of scenarios modeling unwanted user behavior and how that behavior influences the evolution of the dynamical systems, (4) a model of growing activity networks with explicit models of activity potential and peer influence. Summarizing, our results indicate that intrinsic user motivation to

P. Kasper (✉) · P. Koncar · T. Santos · M. Wölbitsch · D. Helic
Graz University of Technology, Graz, Austria
e-mail: patrick.kasper@tugraz.at; philipp.koncar@tugraz.at; teixeiradossantos@tugraz.at;
m.woelbitsch@student.tugraz.at; dhelic@tugraz.at

S. Walk
Detego GmbH, Graz, Austria
e-mail: s.walk@detego.com

M. Strohmaier
RWTH Aachen University, Aachen, Germany
e-mail: markus.strohmaier@humtec.rwth-aachen.de

© Springer Nature Switzerland AG 2019
F. Ghanbarnejad et al. (eds.), *Dynamics On and Of Complex Networks III*,
Springer Proceedings in Complexity, https://doi.org/10.1007/978-3-030-14683-2_5

participate in a collaborative system and peer influence are of primary importance and should be included in the models of the user activity dynamics.

Keywords Nonlinear dynamics · Activity dynamics · Peer influence · Dynamical systems · Collaboration network · Network analysis · Opinion dynamics

1   Introduction

New collaboration websites continuously emerge on the Web. Users of such communities work together towards a defined goal (e.g., building a knowledge base), which sets collaboration websites apart from more common social networks. Whereas some collaboration websites reach a sufficient level of user-activity to sustain themselves, preventing a transition towards inactivity, many websites perish over time or fail to establish an active community at all. The Q&A platform StackOverflow[1] is a successful example of such a collaboration website. Users can ask questions on programming related topics or share their knowledge by answering questions from other members of the community. The explicit goal of the website states *With your help, we're working together to build a library of detailed answers to every question about programming.*[2] A declining community may struggle to meet this ambitious goal in an ever-growing subject field such as programming. Thus, the success of the StackOverflow website relies heavily on the active community collaborating to answer any open questions. However, we as research community still do not fully understand the factors that drive the users to participate and contribute to such websites. This understanding would allow us to support the website operators in their efforts to build a successful website around a flourishing user community.

Initial work in this field frequently concentrated on interactions between users on websites, or how information spreads through the community [1–13]. Nevertheless, to predict success and potentially support websites in their efforts to reach self-sustainability, we argue that understanding as well as modeling the various aspects of user dynamics that go beyond information spreading is of crucial importance.

One of the major problems faced by both new and existing collaboration websites—such as Wikipedia or StackOverflow—revolves around efficiently identifying and motivating the appropriate users to contribute new content. In an optimal scenario, any newly contributed content provides enough incentive on its own, triggering further actions and contributions. Once such a self-reinforced state of increased activity is reached, the system becomes self-sustaining, meaning that sufficiently high levels of activity are reached, which will keep the system active

---

[1] https://stackoverflow.com/.

[2] https://stackoverflow.com/tour.

without further external impulses. StackOverflow is an example for a highly active collaboration website that has already become self-sustained (in terms of activity), evident in the steadily growing number of supporters and overall activity.

However, these self-sustaining states [14–17] are neither easy to reach nor guaranteed to last. For example, Suh et al. [18] showed that the growth of Wikipedia is slowing down, indicating a loss in momentum and perhaps even first evidence of a collapse. Moreover, we generally lack the tools to properly analyze these trends in activity dynamics and thus, cannot even perform tasks such as detecting these self-sustaining system states. Therefore, we argue that new tools and techniques are needed to model, monitor, and simulate the dynamics in collaboration websites.

In this chapter, we set out to shed further light on the complex user dynamics in collaboration websites. More specifically, to investigate the success and failure of collaboration websites, we are interested in the factors that govern growth and decline of the activity in such communities. Moreover, we also aim at evaluating the *robustness* and *stability* of collaborative websites.

Approach To this end, we present a diversified range of approaches, each tackling different aspects of user dynamics in collaboration websites. We use empiric data originating from various types of collaboration websites, such as StackExchange instances and Semantic MediaWikis to report our findings.

We argue that there are two factors that influence the activity of any single user in collaboration websites. First, the activity or rate of contributions of a user is influenced by their intrinsic motivation to participate in a collaborative community. This motivation may decay over time in a mechanism called *activity decay*. A previously active user may lose interest in the community and contribute less and less over time unless stimulated through other means. This behavior has been observed in many different websites [17–19]. In another scenario the intrinsic motivation of a user may remain constant or even increase with time. We summarize this phenomenon as *activity potential*. Second, *peer influence* is a mechanism in which users influence other members of the community. For example, when users post a question to StackExchange and receive helpful answers from other users, they may want to help others in the same way by answering other open questions. Note, contributions by peers are not necessarily always positive. Internet trolls may attempt to disrupt the community by adding detrimental content [20].

We discuss these influential forces and their interactions by (1) applying several tests for nonlinearity on the activity time series of various StackExchange instances to reveal *complex user behavior*. Thereafter, we (2) apply a dynamical systems model to investigate the *long-term activity decay* (users losing interest over time) and how this decay is countered by the peer influence from the other users. Iterating upon this idea of peer influence we (3) conduct experiments investigating the *influence of trolls* who spread negative activity through peer influence by adding detrimental content to the websites, and lastly, we (4) present a generative probabilistic model to create synthetic activity networks and study the *emergence of clustering* in the underlying user networks.

Contribution This chapter provides an overview of several methods and ideas concerning dynamics in collaboration websites. Further, we shed light on some factors contributing to their eventual success or failure. We summarize our main findings as follows. Models incorporating the user-centered concepts of *user motivation* and *peer influence* can capture crucial aspects regarding activity in collaboration websites, such as system robustness and stability. Further, depending on a particular community that we investigate the technical approaches and models need to be carefully chosen.

2   Related Work

Analysis of Online Communities We know that, at some point in time, well-established collaboration websites, such as StackOverflow, have become self-sustained. There, sufficiently high levels of activity are reached, which will keep the system active without further external impulses. However, many websites never reach this state and those that do are not guaranteed to remain there indefinitely [14–17]. With the continuous growth in the number of such websites, many researchers have investigated these communities to better understand the dynamics governing growth and decline. For example, Schoberth et al. [21] and Crandall et al. [22] analyzed time-series data of websites to investigate the communication activities and social influences of their users. Analyzing the roles different types of users play, researchers characterized the users to infer properties about their communities as a whole [23–26]. Using methods related to the work by Zhang et al. [27], multiple authors studied the evolution-dynamics of Web communities and their underlying networks [28–33]. These networks often serve as a basis for dynamical systems models of the communities.

Nonlinear Time Series Analysis To obtain a better understanding of the properties in high-dimensional dynamical systems, researchers have utilized nonlinear time series analysis. Bradley and Kantz [34] provided a thorough overview of applied nonlinear time series analysis. The works by Eckmann et al. [35] and Marwan et al. [36] described the use of recurrence plots to visually analyze complex systems. Zbilut and Webber [37, 38] further extended these visualizations with a method called recurrence quantification analysis (RQA). These tools provided means to, for example, investigate the chaotic behavior in stock markets [39, 40] or predict the outcome of casino games, such as a roulette wheel [41].

Here, we present work employing various tests for nonlinearity to reveal latent nonlinear behavior in collaborative websites and their communities.

Dynamical Systems and Activity Dynamics Dynamical systems in a non-network context are a well-studied scientific and engineering field. Strogatz [42] and Barrat et al. [43] provided an in-depth introduction to dynamical systems. Within the contextual scope of online communities, researchers primarily used dynamical systems to analyze and understand the diffusion of information in online social-networks for purposes such as viral marketing [9–13]. Recently, in the context of activity dynamics, Ribeiro [31] conducted an analysis of the daily number of active users who visit specific websites, fitting a model that allows predicting if a website has reached self-sustainability, defined by the shape of the curve of the daily number of active users over time.

In this chapter, we present a model to simulate activity as a dynamical system on online collaboration networks. Here, two forces, decay of motivation and peer influence govern the activity-potential of users. Moreover, we describe work on how these concepts facilitate the generation of synthetic networks. Online communities becoming increasingly accostable to their users does not always lead to higher overall activity. Internet trolls, for example, generate unwanted content [20, 44–48], creating additional strain for others who attempt to keep the community healthy.

Thus, we present an extension to the previous model incorporating the idea of trolls emitting negative peer influence and discuss how such negative activity can impact the user dynamics in collaboration websites.

3 Datasets

The Web offers a multitude of ways in which people can communicate and collaborate in a group. To capture some of this diversity, we utilize empirical datasets stemming from different types of collaboration websites. Here, we provide a general overview of the empiric datasets in our experiments, and how we extract the user networks from the raw data.

StackExchange instances   StackExchange is a network of currently 172 Question and Answer communities. Here, users can post questions and other members of the community can provide and discuss answers. Some of the most popular instances are StackOverflow[3] and the English StackExchange.[4] We extract the network by representing each user with a node and draw an edge whenever user A replies to a post by user B. The full dataset from which we draw our networks is publicly available.[5] We denote these datasets with an *SE* suffix. For example, we call the network extracted from the English StackExchange as *englishSE*.

Semantic MediaWikis   The Semantic MediaWiki[6] is an extension to the MediaWiki software and allows for storing and querying structured data within the Wiki. We build the community network by representing each contributor with a node and draw an edge whenever two users work on the same page. We collected the data we use in our experiments from the live MediaWiki API, which is now unavailable. However, a comprehensive dump of the Semantic MediaWiki is

---

[3] https://stackoverflow.com/.

[4] https://english.stackexchange.com/.

[5] https://archive.org/details/stackexchange.

[6] https://www.semantic-mediawiki.org.

publicly available.[7] We denote these datasets with an *MW* suffix. For example, we call the network extracted from the Neurolex Semantic MediaWiki as *neurolexMW*.

SubReddits   A SubReddit is a community within Reddit for a specific topic. While some of these communities act as recommendation platforms or Q&A sites akin to StackExchange, others aim to facilitate a platform for open discussion of various topics. We extract a network from a SubReddit by representing each user with a node and draw an edge when one user replies to a post by another user. These dumps from Reddit are publicly available.[8] We denote these datasets with an *SR* suffix. For example, we call the network extracted from the Star Wars Subreddit as *starwarsSR*.

4   Complex User Behavior in Collaboration Websites

As a first step towards the goal of identifying factors indicating successful or failing collaboration websites, we set out to identify complex (nonlinear) user behavior present in the data. To reveal and characterize any hidden nonlinear patterns, we construct the activity time series from the datasets of 16 randomly selected StackExchange instances and conduct a set of nine established tests for nonlinearity on them. This information allows for a decision on whether a standard time-series model such as the autoregressive integrated moving average (ARIMA) is sufficient to capture and predict activity or more complex approaches (e.g., dynamical systems) should be employed.

Activity Time Series   We construct the activity time series from a dataset by first measuring the activity—the number of questions, answers, and comments—per day. To remove outliers in the data we smooth the time series with a rolling mean over a 7-day period. Finally, we calculate the sum of the smoothed activity over all users per week, yielding a time series with one entry per week representing the activity in the corresponding community.

Experiments and Results   To reveal hidden nonlinear patterns in our activity time series, we apply the following tests for nonlinearity on each dataset and report the results: (i) *Broock, Dechert, and Scheinkman test* [49]; (ii) *Teraesvirta neural network test* [50]; (iii) *White neural network test* [51]; (iv) *Keenan one-degree test for nonlinearity* [52]; (v) *McLeod–Li test* [53]; (vi) *Tsay test for nonlinearity* [54]; (vii) *Likelihood ratio test for threshold nonlinearity* [55]; (viii) *Wald–Wolfowitz runs test* [55, 56]; (ix) *Surrogate test–time asymmetry* [57].

We apply these tests without configuration changes, except for the *Broock, Dechert, and Scheinkman* and *Wald–Wolfowitz runs* tests. As described in Zivot and

---

[7]https://archive.org/details/wiki-neurolexorg_w.

[8]https://files.pushshift.io/reddit/.

Wang [58, p. 652], we compute the test statistic of *Broock, Dechert, and Scheinkman* on the residuals of an ARIMA model, to check for nonlinearity not captured by ARIMA. For the *Wald–Wolfowitz runs* test, since a run represents a series of similar responses, we define a positive run as the number of times the time series value was greater than the previous one [59].

To validate the plausibility of this categorization we compare the forecast performance from three standard time series models, namely ARIMA, exponential smoothing models (ETS), and linear regression models, with nonlinear models, reconstructed from the observed activity time series.

Table 1 lists test results on the 16 StackExchange instances. Our results reveal that on the one hand, there are StackExchange communities with mostly linear behavior, such as *englishSE* and *unixSE* as only two tests suggest nonlinearity. On the other, we see that for the communities *bicycleSE*, *bitcoinSE*, and *mathSE* the majority of tests suggest nonlinearity.

A higher number of tests suggesting nonlinearity for a community indicates a better fit for models based on nonlinear time-series analysis. The prediction experiments and the Friedman test ranks [60] on datasets with mostly negative test results (less than five) indicate that for these communities ARIMA and ETS models result in the best fit. For the other datasets (more than four positive tests), nonlinear models yield the lowest error.

The nonlinearity tests by Lee et al. [51] and Teräsvirta et al. [50] utilize neural networks and appear to be more sensitive to the presence of nonlinear dynamics than the other tests, since they test positive for nonlinearity four times more often in the dataset group with five or more tests indicating nonlinearity than in the other dataset group. We attribute the usefulness of these two tests to the well-studied ability of neural networks to model nonlinear behavior.

In a second experiment, we use with recurrence plots [36] to analyze the nonlinear properties for two exemplary StackExchange instances *bitcoinSE* and *mathSE*. Both websites have a high number of positive nonlinearity tests.

Figure 1 illustrates the results for these two instances. Despite having the same number of positive tests for nonlinearity, these visualizations depict different patterns in their activity. In particular, Fig. 1b shows a higher density of recurrence points in the upper left corner, gradually diminishing towards the lower right corner. This structure reveals a drift pattern which is present even after linear detrending.

Findings We find that we can model activity on collaboration websites through reconstruction of their underlying, dynamical systems, with some communities showing more signs of nonlinear behavior than others. In particular, the knowledge of any drift- or periodicity patterns in the data provides information on which approach may yield the best accuracy.

For a more detailed discussion of the topic, refer to Santos et al. [61].

Table 1 Results of statistical tests

Dataset	Weeks	τ	m	Nonlinearity test score	Positive nonlinearity tests	RMSE			
						ARIMA	ETS	Linear	Nonlinear
englishSE[b]	240	2	9	2/9	(i),(v)	0.679	0.445	0.332	**0.308**
unixSE[b]	239	1	7	2/9	(i),(v)	0.209	0.209	0.241	**0.207**
chemistrySE[b]	158	2	7	3/9	(i),(v),(viii)	0.498	**0.253**	0.324	0.461
webmastersSE	244	1	8	3/9	(iv),(v),(ix)	**0.231**	0.252	0.334	0.234
chessSE	148	2	8	4/9	(i),(iv),(v),(ix)	**0.254**	–[a]	0.562	0.511
historySE	177	1	9	4/9	(i),(iv),(v),(viii)	0.350	**0.236**	0.304	0.405
linguisticsSE	181	2	6	4/9	(i),(iv),(v),(ix)	**0.251**	0.270	0.300	0.328
sqaSE	200	3	9	4/9	(i),(iv),(v),(ix)	1.813	**0.253**	0.654	0.390
texSE[b]	241	1	7	4/9	(v),(vi),(viii),(ix)	**0.158**	**0.158**	0.276	0.275
tridionSE	107	1	7	4/9	(ii),(iii),(iv),(v)	**0.271**	–[a]	0.614	–[a]
Friedman test rank on datasets with nonlinearity test score < 5/9						2	**1**	4	3
arduinoSE	56	1	10	5/9	(i),(ii),(iii),(iv),(v)	**0.348**	–[a]	–[a]	–[a]
sportsSE	159	1	7	5/9	(i),(iv),(v),(viii),(ix)	**0.244**	0.334	0.401	0.332
uxSE	239	2	8	5/9	(i),(iii),(iv),(v),(vi)	0.347	0.174	0.349	**0.137**
bitcoinSE	182	4	11	6/9	(i),(ii),(iii),(iv),(v),(ix)	0.609	0.554	0.593	**0.578**
mathSE[b]	242	2	8	6/9	(i),(ii),(v),(vi),(viii),(ix)	**0.132**	0.231	0.352	0.291
bicyclesSE	235	2	7	7/9	(i),(ii),(iii),(iv),(v),(viii),(ix)	0.297	0.309	0.325	**0.280**
Friedman test rank on datasets with nonlinearity test score ≥ 5/9						2[c]	2[c]	4	1

This table lists the activity time series length in weeks, embedding parameters τ and m, the number and reference of statistical tests indicating nonlinearity ($\alpha = 0.05$), and the RMSE (lower is better) of a 1 year forecast per model for each dataset. Further it lists the ranking the Friedman test for datasets with less than or more tests suggesting nonlinearity. The indices refer the individual tests as listed in Sect. 4

[a] This activity time series is too short for a 1 year forecast with this model

[b] This activity time series had a strong linear trend, so the results above concern the activity time series detrended with linear regression

[c] These models achieved the same rank in the Friedman test for this group of datasets

Bold values indicate the best RMSE score across the four tested models

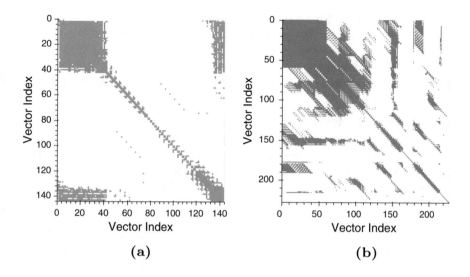

Fig. 1 Recurrence plots (RP) for activity time series
This figure illustrates the recurrence plots of the (**a**) bitcoinSE and (**b**) mathSE websites. Figure (**b**) shows a higher density of recurrence points in the upper left corner, gradually diminishing towards the lower right; this is a sign of a drift in the activity time series, still present after removing the linear trend. Both examples hint at non-stationary transitions in the activity time series

5 Activity Decay and Peer Influence

On collaboration websites contributing users tend to lose interest over time. Wikipedia is a prominent example of such a website with a declining user-base [19]. To address this problem, we present a model based on dynamical systems where the motivation of a user decays over time (intrinsic activity decay). Danescu-Niculescu-Mizil et al. [25] were able to observe this behavior across different online communities. However, in our proposed model, users also gain activity from their neighbors through peer-influence to compensate for the intrinsic decay, which builds upon the notion that people tend to copy their friends and peers [62–64]. This activity dynamics model is capable of capturing and simulating activity in collaboration websites. We fit this model to a number of StackExchange instances and Semantic MediaWikis to simulate trends in activity dynamics. Further, we utilize the model to calculate a threshold indicating self-sustainability. Being able to monitor and measure the stability of a website with regard to user activity indicates how susceptible a system is to fluctuating members. For example, in a volatile website, a small number of highly active users (emitting a lot of peer influence) leaving could result in activity decreasing to the point of total inactivity.

Dynamical Systems The proposed model utilizes the formalism of dynamical systems—meaning that activity is modeled by a system of coupled nonlinear differential equations. Each user in the system is represented by a single quantity (the current activity), and the collaborative ties between users define the coupling of variables.

The model builds on two mechanisms which postulate that with time users lose interest to contribute and that, on the other hand, users are influenced by the actions taken by their peers.

Modeling Activity We model activity dynamics in an online collaboration network as a dynamical system on a network. Hereby, the nodes of a network represent users of the system and links represent the fact that the users have collaborated in the past. We represent the network with an $n \times n$ adjacency matrix A, where n is the number of nodes (users) in the network. We set $A_{ij} = 1$ if nodes i and j are connected by a link and $A_{ij} = 0$ otherwise. Since collaboration links are undirected, the matrix A is symmetric, thus $A_{ij} = A_{ji}$, for all i and j.

We model activity as a continuous real-valued dimensionless variable x_i (representing ratio of the current activity of user i over some critical activity threshold) evolving on node i of the network in continuous dimensionless time τ. We write the time evolution equation as follows:

$$\frac{dx_i}{d\tau} = -\frac{\lambda}{\mu}x_i + \sum_j A_{ij} \frac{x_j}{\sqrt{1 + x_j^2}}. \tag{1}$$

There is only one parameter in our dynamics equation, namely the ratio λ/μ. This is a dimensionless ratio of two rates: (1) The *Activity Decay Rate* λ, which is the rate at which a user loses activity (or motivation), and (2) the *Peer Influence Growth Rate* μ, which is the rate at which a user gains activity due to the influence of a *single* neighbor.

The ratio between those two rates is the ratio of how much faster users lose activity due to the decay of motivation than they can gain due to positive peer influence of a single neighbor. For example, a ratio of $\lambda/\mu = 100$ would mean that the users intrinsically lose activity 100 times faster than they potentially can get back from one of their neighbors.

The master stability equation for our activity dynamics model is

$$\kappa_1 < \frac{\lambda}{\mu}, \tag{2}$$

where κ_1 is the largest positive eigenvalue of the graph adjacency matrix. Note that this inequality separates the network structure (κ_1) from the activity dynamics (λ/μ). If this stability condition is satisfied, the fixed point $x^* = 0$, in which there is no activity at all ("inactive" system), represents a stable fixed point. This also means that small changes in activity only cause the system to momentarily leave the (attracting) fixed point until it becomes inactive again.

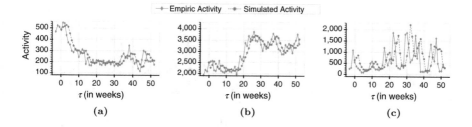

Fig. 2 Activity simulation
The figure depicts the results of our activity dynamics simulation for the StackExchange datasets and Semantic MediaWikis. In all our analyzed datasets, the simulated activity dynamics exhibit a notable resemblance to the empirical activity. (**a**) bitcoinSE. (**b**) englishSE. (**c**) neurolexMW

Experiments and Results To estimate λ/μ for the empirical datasets we employ an output-error estimation method. First, we formulate the estimation of the model parameter as an optimization problem. As objective function, we use a least-squares cost function. Second, we solve the optimization problem numerically, using the method of gradient descent in combination with the Newton–Raphson method [65] to speed up the calculations. Finally, we evaluate the accuracy of the ratio estimate by calculating prediction errors on unseen data.

This prediction serves as a demonstration that our assumptions regarding the *Activity Decay Rate* and the *Peer Influence Growth Rate* hold and allow us to simulate trends in activity dynamics for given and real values. The simplifications, such as the static network structure and average model parameters over weeks and users, entail that any results cannot be used for an accurate prediction of the activity in the system, and naturally limit the accuracy of our results. These limitations are particularly visible whenever there are large and sudden increases in activity in the collaboration websites. Figure 2 depicts the results of the activity dynamics simulation. Overall, the results gathered from the activity dynamics simulation exhibit notable resemblance to the real activities of the corresponding datasets. Note how in some cases our simulation yields a higher activity increase than the real data (e.g., Fig. 2c). A possible cause for this behavior is the static network structure where users might be influenced by peers who actually join the network at a later point in time.

Figure 3 depicts the value of the calculated ratios λ/μ (y-axis) for each week (x-axis) of our activity dynamics simulation. If the ratio is higher than κ_1, our master stability equation holds, and the system converges towards zero activity (over time). The amount of activity that is lost per iteration—and hence the speed of activity loss—is proportional to the value of the ratio and the activity already present in the network. In general, a higher ratio results in a higher and faster loss of activity.

If the ratio is smaller than κ_1, the master stability equation has been invalidated and the system will converge towards a new fixed point of immanent activity (cf. Eq. 2). Robust systems are lively and high levels of activity, which are able to

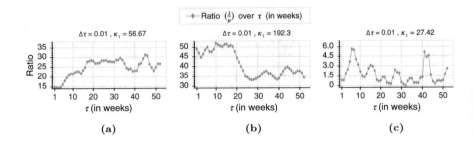

Fig. 3 Evolution of ratios λ/μ
The evolution of the ratios λ/μ (y-axes) over τ (in weeks; x-axes) for the StackExchange datasets and for the Semantic MediaWikis. The smaller the ratio, the higher the levels of activity in Fig. 2. Small variances in λ/μ over time indicate that activities of the systems are less influenced by the activity of single individuals than they are by peer influence. (**a**) bitcoinSE. (**b**) englishSE. (**c**) neurolexMW

keep that activity even in the cases of small unfavorable changes in the dynamical parameters.

Note that one advantage of our model over other existing approaches, such as autoregression, is the interpretability of the ratio λ/μ. For example, a ratio of 4 means that users intrinsically lose activity four times faster than they can get back from one of their peers, while the coefficients of the autoregression lack such interpretable characteristics. Further, using the concept of dynamical systems we can represent the underlying mechanisms in a closed form. This allows for further detailed analytical inspections, such as a linear stability analysis, which is much harder, if not impossible, to conduct for other models (e.g., agent-based models, autoregression or more complex models based on dynamical systems).

Practical Implications Using our proposed model, we can characterize networks based on their susceptibility to changes in activity (referred to as Activity Momentum in [32]). Hence, community managers could use the proposed model as indicator for the robustness of their collaboration website with regard to its activity dynamics.

Further, we can characterize the potential of a collaboration network to become self-sustaining by comparing the calculated ratios of λ/μ with the corresponding κ_1 and the susceptibility to changes in user activity of the collaboration network. If the ratio is below κ_1, our master stability equation is invalidated, pushing the system towards a new fixed point where the forces of the Activity Decay Rate and the Peer Influence Growth Rate reach an equilibrium so that the network converges towards a state of immanent and lasting activity. If such a state is reached combined with a low susceptibility to changes in user activity, the corresponding collaboration network has reached critical mass of activity and has become self-sustaining; no external impulses are required to keep the network active.

Of course, in real-world scenarios, activity will not last forever without providing additional incentives (e.g., user profile badges displaying support or expertise), as interest (and thus activity) in a system potentially decays over time. As a consequence, this would first result in an increase of μ and inevitably, with a sufficiently large μ, the collaboration network would return to its stable fixed point, once our master stability equation holds again, and activity would once more converge towards zero.

Findings Using our proposed model to simulate activity dynamics, we show that the overall activity in collaboration websites appears to be a composite of the *Activity Decay Rate* and the *Peer Influence Growth Rate*. A first analysis of the model suggests that activity dynamics in collaboration networks have an obvious and natural fixed point—the point of complete inactivity—where all contributions of the users have seized. However, by slightly manipulating the parameters in our model we show that it is possible to destabilize the fixed point, resulting in a potential increase in activity.

For a more detailed presentation and discussion of factors such as *system mass* and *Activity Momentum*, see Walk et al. [32].

6 Negative Activity in Collaboration Websites

While most users in collaboration networks contribute by adding helpful content— in the case of StackExchange by asking questions or providing helpful answers— Internet Trolls post unwanted content for their own amusement [20]. We investigate such unwanted users and how they affect collaboration websites by adapting the activity dynamics model presented in Sect. 5. Thus far, we considered peer influence as a purely positive force. In this proposed modification, introduced trolls emit negative activity to their neighbors. As an example, a troll may post a nonsensical question on StackExchange, or deliberately post wrong answers. Other users now have to spend time to either report or remove the unwanted post. We argue that this consumes the time these users could have potentially used to answer an open question. Understanding how trolls can disrupt the activity in collaboration websites can be used to derive strategies to prevent or minimize their impact.

Modeling Troll-Users We model the impact of disruptive content in the form of negative activity. A troll-user emits negative activity to their connected users and simultaneously receive productive positive activity as their neighbors try to compensate for it. Further, we argue that trolls commit to their cause and therefore do not lose motivation on their own. Thus, we disable the *motivation decay* for these users. Within a network, we define the total number of normal users as N and the number of trolls as T. Thus, N remains constant regardless of how many trolls enter the network.

Whenever a troll enters a network at the beginning of our experiments, they connect to a number of existing users (α). We define two methods for this process; First, with the *random strategy* the troll connects to other users uniformly at random ($P = \frac{\alpha}{N}$). To achieve this, the troll may extract a list of all users within the network

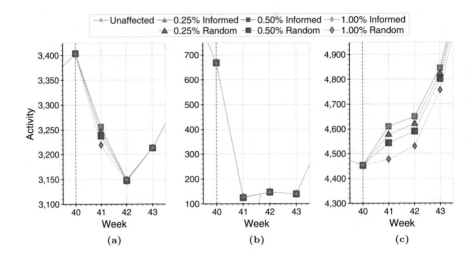

Fig. 4 Effects of trolls on total system activity
This figure depicts the impact trolls have for the first weeks after their introduction to the network. When trolls connect to highly active users (informed strategy) the effect on the systems activity is minimal, whereas with the random connection strategy we observe a noticeable impact on the activity in the network. (**a**) englishSE. (**b**) neurolexMW. (**c**) starwarsSR

and then perform a random selection. Second, with the *informed strategy* the troll specifically targets and connects to highest degree users. Here the troll observes the collaboration website for some time before selecting their targets according to this strategy.

The negative activity of a troll absorbs the positive activity spread via peer influence. Note that, when a normal user receives enough negative activity, their own activity can become negative for some time. Whenever the incoming peer influence received by a troll exceeds their outgoing activity, the troll is defeated, and we remove their corresponding node from the network.

Experiments & Results In this experiment, we aim to determine how trolls affect the overall activity in networks. For the initial 39 weeks we calculate the activity within a network unaltered (akin to the model in Sect. 5). After this point, we introduce the troll-users and observe their impact. For each troll, we set their starting activity at week 40 to -5 and conduct each experiment twice. Once, with trolls following the random strategy, and once with informed connection strategy. Further, we fix the parameter α (number of connections per troll) to be equal to the mean degree of all existing users. In total, we add trolls equal to the amount of 0.25%, 0.50%, and 1.00% of existing users (N) and investigate their initial impact.

Figure 4 illustrates the simulated system activity for the first 3 weeks after we add the trolls to the networks. Our results suggest that trolls connecting to highly active users do not affect the overall activity in the network. We attribute this to the peer influence emitted by the troll being comparably insignificant. However, when we connect the trolls at random, users are more heavily influenced. A sporadic contributor may lose interest upon exposure to trolls. Small, but well-connected networks may lack a sizable body of casual users. Figure 4b illustrates such a network (neurolexMW). Due to this strong structure, newly introduced trolls fail to disrupt the system regardless of their connection strategy.

Findings Up to a threshold, highly active users can compensate for the negative activity trolls emit, whereas random users can be more susceptible and may even temporarily spread negative content on their own further reducing the activity in a network. Below this threshold, the negative activity is nullified over time. However, once enough trolls connect to the highly active users within a network and overwhelm their positive activity, the networks collapse rapidly, ending all productive contribution.

Based on these findings, website administrators may entrust highly active users to moderate their communities. These moderators would be instructed on how to deal properly with occurring trolls, making them less susceptible for distractions (unproductive activity). Additionally, moderators could support peripheral users targeted by trolls and handle other detrimental factors, such as spam bots or illegal content. Offering incentives, such as additional functionality on their website or even money, could help motivating users to become moderators. A different approach would be to use machine learning techniques to automatically detect occurring trolls, for example, by identifying fake profiles [66] or by inspecting textual contents of comments [67, 68].

For further discussion of this subject and experiments on how trolls affect and infect users, see Koncar et al. [69].

7 Peer Influence in Temporal Networks

Thus far, we have represented the collaboration websites and their user networks in a static form. However, in the real world, new users frequently join the communities, while other members leave after a while. We approach this dynamic user-base by presenting a generative model to create synthetic networks. Existing network generators incorporating the concept of activity often solely consider the intrinsic activity potentials as sources of activity [70, 71]. But, we have shown in the previous sections that interaction between users is an important factor to consider. Thus, we present a generative model that incorporates peer influence (similar to Sect. 5) and tie strength (how frequent two users interact) as explicit mechanics. With this model for generating synthetic networks we are able to explore new ideas and conduct experiments before verifying them on empiric networks.

Generating Activity Networks We model the influence that a node receives from their neighbors in each time step (iteration t) as the increase in the activity potential according to the number of active neighbors in the previous iteration ($t - 1$) and the tie strengths.

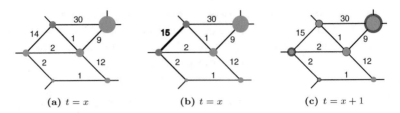

Fig. 5 Illustrative model example
The highlighted node in (**a**) becomes active and interacts with the second highlighted node in (**b**), reinforcing the tie between them. The outlines in (**c**) depict the additional peer-influenced activation probability in the next iteration

The equation for the peer influence for a node v_i is:

$$p_i(t) = \frac{\alpha_i(t)\, q}{\sqrt{\alpha_i^2(t) + \theta^2}}, \tag{3}$$

where $\alpha_i(t)$ is the weighted fraction of active neighbors and q is the parameter for the maximum peer influence. Further, $(\theta > 0)$ denotes a critical threshold, determining the required fraction of active neighbors to set the peer influence probability close to this maximum.

Any node can become active based on either their own intrinsic activity or on the peer influence. When they do they select a new node as the partner for the interaction and either create or reinforce the tie between them.

The resulting network exhibits structures seen in real-world networks, due to the partner selection, which follows a set of rules. First, a memory effect as described by Karsai et al. [72] (depending on the number of currently existing ties for the node and the memory strength parameter c) defines the probability to reinforce an existing tie. More precisely, this probability is equal to $\frac{c}{k_i + c}$, where k_i is the number of current neighbors. Second, if a node wants to form a new tie, it tries to perform a cyclic closure [73]—by interacting with a randomly selected neighbor of a neighbor—with the probability $p_\triangle$, or a focal closure [74], which emulates homophily (i.e., similar users connect to each other). The latter is performed with a probability of $1 - p_\triangle$, or if there are no suitable candidates for a cyclical closure. This is, for example, the case if a node becomes active for the first time.

Figure 5 illustrates these mechanics. Figure 5a describes the network at iteration step $t = x$, where the numbers along the edges represent the tie strengths and the site of the nodes indicates their intrinsic activity potential. In this example, the highlighted node (top left) becomes active. It selects the newly highlighted node (left) in Fig. 5b as the partner, which becomes active as due to this interaction. As a result, they reinforce the tie between them. At the start of the next iteration, the nodes receive peer influence from their neighbors active in the last iteration (outlines in Fig. 5b). Note how the node in the top right corner receives a high amount of peer influence due to its strong ties.

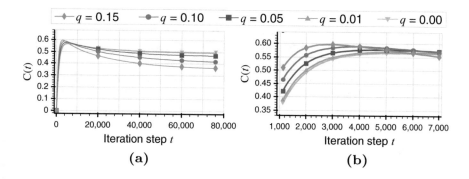

Fig. 6 **Average clustering coefficient ($C(t)$) evolution**
This figure depicts the average clustering coefficient (*y*-axis) at each iteration step (*x*-axis) over various values for the maximum peer influence (q). Higher values for q result in stronger peer influence effects. Note that the value of q also affects the time until convergence. Figure (**b**) illustrates the timespan where the $C(t)$ is maximal. (**a**) Visualizes the full simulation whereas (**b**) is a cutout where $C(t)$ is maximal

To prevent the network from becoming fully connected after a sufficient amount of iterations, every node has a probability to be removed. In this case, we delete the node from the network and introduce a new node (without any existing ties). As a result, the total number of nodes in the network remains constant.

Experiments and Results We generate synthetic networks with 5000 nodes over 75,000 iterations with varying values for the maximum peer influence parameter (q). To ensure the formation of adequate community structures in the network we set $p_\Delta = 0.9$ and the probability for node-deletion to $p_d = 5 * 10^{-5}$. Further, we fix the parameter for memory strength to favor new ties ($c = 1$) and fix the critical peer influence threshold to $\theta = 0.1$ to reflect the intuition that a small number of active neighbors is sufficient to affect the activity of a node to a large extent. Finally, we run each configuration 40 times to account for statistical fluctuations and report average results.

Figure 6 illustrates these results. For the first few hundred iterations, the clustering coefficient ($C(t)$) is low but rapidly increases until it reaches its maximum between iteration $t = 3000$ and $t = 5000$. After this peak, it slowly declines until the network eventually reaches a stable state. Further, higher values for q increase the speed at which the maximum is reached but also result in a lower average clustering coefficient once the network is stable. As the peer influence mechanism increases the activity in the network, especially in already formed communities, increases and active nodes motivate their neighbors to become more active.

Findings Peer influence is an effective mechanism for the creation of synthetic activity networks. We present a model creating networks that exhibit similar community structures to real-world networks, such as triadic closures (three users all connected with each other) [70]. Further, we show that during the first few iterations the average clustering coefficient increases, indicating that during the early stages of a network, activity is concentrated on a core of highly active users. After reaching a peak activity starts to spread out more evenly throughout the system, indicated by a slow and steady decline of the average clustering coefficient.

For further details and an analysis of inter-event time distributions (*burstiness*), see the full paper on the topic by Wölbitsch et al. [75].

8 Conclusions

In this chapter, we asked the overarching question what factors govern growth and decline of activity in collaboration websites and how to evaluate their robustness and stability.

To this end, we presented and discussed various approaches to investigate a range of aspects influencing the user dynamics in collaboration websites. First, we used tests to assess the presence of complex user behavior by analyzing the nonlinearity of activity time series. Second, we presented a model based on dynamical systems, incorporating the concepts of loss of motivation (activity decay) and users affecting their neighbors (peer influence) to model and simulate activity in a collaboration website. Third, we introduced a modification to this model to simulate the impact of trolls (spreading negative peer influence). Fourth, we utilized activity potentials and peer influence in a generative model to create synthetic activity networks. Collectively, we summarize our key findings as follows.

Complex user behavior   Our results suggest that user activity varies across different collaboration websites with some communities exhibiting more signs of nonlinear behavior than others.

Activity Decay and Peer influence   We find that intrinsic activity decay and peer influence serve as viable mechanics to capture and simulate activity in collaboration websites. Further, we can employ this peer influence to investigate the impact of troll-users on a system.

Activity Potentials   Lastly, we can extend the concept of user motivation through the mechanism of activity potentials and utilize this concept in combination with peer influence to generate synthetic activity networks that exhibit structures also present in their real-world counterparts.

The work we present in this chapter extends the body of existing research on dynamics in collaboration websites and may serve as a base for further research to predict the eventual success or failure of a collaboration website at an early stage. Finally, we demonstrated how the viability of an approach to analyze user dynamics in collaboration websites depends on the investigated aspect and the information available in the data.

References

1. Zang, C., Cui, P., Faloutsos, C., Zhu, W. In: Proceedings of the 23rd ACM SIGKDD International Conference on Knowledge Discovery and Data Mining, pp. 565–574. ACM, New York (2017). https://doi.org/10.1145/3097983.3098055
2. Baronchelli, A., Felici, M., Loreto, V., Caglioti, E., Steels, L.: J. Stat. Mech: Theory Exp. **2006**(06), P06014 (2006). https://doi.org/10.1088/1742-5468/2006/06/P06014
3. Sznajd-Weron, K., Sznajd, J.: Int. J. Mod. Phys. C **11**(06), 1157 (2000). https://doi.org/10.1142/S0129183100000936
4. Krapivsky, P.L., Redner, S.: Phys. Rev. Lett. **90**(23), 238701 (2003). https://doi.org/10.1103/PhysRevLett.90.238701
5. Martins, A.C.: Int. J. Mod. Phys. C **19**(04), 617 (2008). https://doi.org/10.1142/S0129183108012339
6. De, A., Valera, I., Ganguly, N., Bhattacharya, S., Gomez-Rodriguez, M. In: Proceedings of the 30th International Conference on Neural Information Processing Systems, pp. 397–405. Curran Associates Inc., Red Hook (2016)
7. Fan, K., Pedrycz, W.: Physica A **462**, 431 (2016). https://doi.org/10.1016/j.physa.2016.06.110
8. Wang, C. (2016): An Opinion Dynamics Model with Increasing Self-Confidence. Preprint. arXiv:1609.05732
9. Leskovec, J., Adamic, L.A., Huberman, B.A.: ACM Trans. Web **1**(1), 5 (2007). https://doi.org/10.1145/1232722.1232727
10. Leskovec, J., Backstrom, L., Kleinberg, J. In: Proceedings of the 15th ACM SIGKDD International Conference on Knowledge Discovery and Data Mining, pp. 497–506. ACM, New York (2009). https://doi.org/10.1145/1557019.1557077
11. Myers, S.A., Zhu, C., Leskovec, J. In: Proceedings of the 18th ACM SIGKDD International Conference on Knowledge Discovery and Data Mining, pp. 33–41. ACM, New York (2012). https://doi.org/10.1145/2339530.2339540
12. Vespignani, A.: Nat. Phys. **8**(1), 32 (2012). https://doi.org/10.1038/nphys2160
13. Iribarren, J.L., Moro, E.: Phys. Rev. Lett. **103**, 038702 (2009). https://doi.org/10.1103/PhysRevLett.103.038702
14. Oliver, P., Marwell, G., Teixeira, R.: Am. J. Sociol. **91**, 522–556 (1985). https://doi.org/10.1086/228313
15. Oliver, P.E., Marwell, G.: Am. Sociol. Rev. **54**, 1–8 (1988). https://doi.org/10.2307/2095728
16. Marwell, G., Oliver, P.E., Prahl, R.: Am. J. Sociol. **94**(3), 502 (1988). https://doi.org/10.1086/229028
17. Solomon, J., Wash, R. In: Proceedings of the International AAAI Conference on Weblogs and Social Media (ICWSM) (2014)
18. Suh, B., Convertino, G., Chi, E.H., Pirolli, P. In: WikiSym '09: Proceedings of the 5th International Symposium on Wikis and Open Collaboration, pp. 1–10. ACM, Orlando (2009). https://doi.org/10.1145/1641309.1641322
19. Halfaker, A., Geiger, R.S., Morgan, J.T., Riedl, J.: Am. Behav. Sci. **57**(5), 664 (2013). https://doi.org/10.1177/0002764212469365
20. Shin, J. In: Society for Information Technology & Teacher Education International Conference, pp. 2834–2840. Association for the Advancement of Computing in Education (AACE), Waynesville (2008)
21. Schoberth, T., Preece, J., Heinzl, A. In: Proceedings of the 36th Annual Hawaii International Conference On System Sciences, 10 pp. IEEE, Piscataway (2003). https://doi.org/10.1109/HICSS.2003.1174576
22. Crandall, D., Cosley, D., Huttenlocher, D., Kleinberg, J., Suri, S. In: Proceedings of the 14th ACM SIGKDD International Conference on Knowledge Discovery and Data Mining, pp. 160–168. ACM, New York (2008). https://doi.org/10.1145/1401890.1401914
23. Mamykina, L., Manoim, B., Mittal, M., Hripcsak, G., Hartmann, B. In: Proceedings of the SIGCHI Conference on Human Factors in Computing Systems, pp. 2857–2866. ACM, New York (2011). https://doi.org/10.1145/1978942.1979366

24. Furtado, A., Andrade, N., Oliveira, N., Brasileiro, F. In: Proceedings of the 2013 Conference on Computer Supported Cooperative Work, pp. 1237–1252. ACM, New York (2013). https://doi.org/10.1145/2441776.2441916

25. Danescu-Niculescu-Mizil, C., West, R., Jurafsky, D., Leskovec, J., Potts, C. In: Proceedings of the 22Nd International Conference on World Wide Web, Rio de Janeiro, Brazil, 2013, WWW '13, pp. 307–318. https://doi.org/10.1145/2488388.2488416

26. Yang, J., Tao, K., Bozzon, A., Houben, G.J.: User Modeling, Adaptation, and Personalization, pp. 266–277. Springer, Cham (2014). https://doi.org/10.1007/978-3-319-08786-3_23

27. Zhang, J., Ackerman, M.S., Adamic, L. In: Proceedings of the 16th International Conference on World Wide Web, pp. 221–230. ACM, New York (2007). https://doi.org/10.1145/1242572.1242603

28. Wang, G., Gill, K., Mohanlal, M., Zheng, H., Zhao, B.Y. In: Proceedings of the 22nd International Conference on World Wide Web, pp. 1341–1352. ACM, New York (2013). https://doi.org/10.1145/2488388.2488506

29. Anderson, R.M., May, R.M.: Infectious Diseases of Humans: Dynamics and Control. Oxford University Press, Oxford (1991)

30. Matsubara, Y., Sakurai, Y., Prakash, B.A., Li, L., Faloutsos, C. In: Proceedings of the 18th ACM SIGKDD International Conference on Knowledge Discovery and Data Mining, pp. 6–14. ACM, New York (2012). https://doi.org/10.1145/2339530.2339537

31. Ribeiro, B. In: Proceedings of the 23rd International Conference on World Wide Web, Seoul, Korea, 2014, WWW '14, pp. 653–664. https://doi.org/10.1145/2566486.2567984

32. Walk, S., Helic, D., Geigl, F., Strohmaier, M.: ACM Trans. Web **10**(2), 11 (2016). https://doi.org/10.1145/2873060

33. Zang, C., Cui, P., Faloutsos, C. In: Proceedings of the 22nd ACM SIGKDD International Conference on Knowledge Discovery and Data Mining, pp. 2015–2024. ACM, New York (2016). https://doi.org/10.1145/2939672.2939825

34. Bradley, E., Kantz, H., Chaos: Interdiscip. J. Nonlinear Sci. **25**(9), 097610 (2015). https://doi.org/10.1063/1.4917289

35. Eckmann, J.P., Kamphorst, S.O., Ruelle, D.: Europhys. Lett. **4**(9), 973 (1987). https://doi.org/10.1209/0295-5075/4/9/004

36. Marwan, N., Romano, M.C., Thiel, M., Kurths, J.: Phys. Rep. **438**(5), 237 (2007). https://doi.org/0.1016/j.physrep.2006.11.001

37. Zbilut, J.P., Webber Jr., C.L.: Phys. Lett. A **171**(3–4), 199 (1992). https://doi.org/10.1016/0375-9601(92)90426-M

38. Webber Jr., C.L., Zbilut, J.P.: J. Appl. Physiol. **76**(2), 965 (1994). https://doi.org/10.1152/jappl.1994.76.2.965

39. Hsieh, D.A.: J. Financ. **46**(5), 1839 (1991). https://doi.org/10.1111/j.1540-6261.1991.tb04646.x

40. Strozzi, F., Zaldívar, J.M., Zbilut, J.P.: Physica A **312**(3), 520 (2002). https://doi.org/10.1016/S0378-4371(02)00846-4

41. Small, M., Tse, C.K.: Chaos: Interdiscip. J. Nonlinear Sci. **22**(3), 033150 (2012). https://doi.org/10.1063/1.4753920

42. Strogatz, S.H.: Nonlinear Dynamics and Chaos: With Applications To Physics, Biology, Chemistry, and Engineering (Studies in Nonlinearity). Studies in Nonlinearity. Perseus Books Group, New York (1994)

43. Barrat, A., Barthelemy, M., Vespignani, A.: Dynamical Processes on Complex Networks, vol. 1. Cambridge University Press, Cambridge (2008)

44. Allen, M.: Don't be a troll! Using the Internet for successful higher education. Presented at the Higher Education Conference, Sydney. Curtin University of Technology (1999)

45. Hardaker, C.: Trolling in asynchronous computer-mediated communication: from user discussions to academic definitions. J. Politeness Res. **6**, 215–242 (2010). https://doi.org/10.1515/jplr.2010.011

46. Bergstrom, K.: First Monday **16**(8) (2011). https://doi.org/10.5210/fm.v16i8.3498

47. Buckels, E.E., Trapnell, P.D., Paulhus, D.L.: Personal. Individ. Differ. **67**, 97 (2014). https://doi.org/10.1016/j.paid.2014.01.016

48. Pasquale, F.: How to tame an Internet troll. Chron. High. Educ. **62**(05) (2015)
49. Broock, W., Scheinkman, J.A., Dechert, W.D., LeBaron, B.: Econ. Rev. **15**(3), 197 (1996). https://doi.org/10.1080/07474939608800353
50. Teräsvirta, T., Lin, C.F., Granger, C.W.: J. Time Ser. Anal. **14**(2), 209 (1993). https://doi.org/10.1111/j.1467-9892.1993.tb00139.x
51. Lee, T.H., White, H., Granger, C.W.: J. Econ. **56**(3), 269 (1993). https://doi.org/10.1016/0304-4076(93)90122-L
52. Keenan, D.M.: Biometrika **72**(1), 39 (1985). https://doi.org/10.1093/biomet/72.1.39
53. McLeod, A.I., Li, W.K.: J. Time Ser. Anal. **4**(4), 269 (1983). https://doi.org/10.1111/j.1467-9892.1983.tb00373.x
54. Tsay, R.S.: Biometrika **73**(2), 461 (1986). https://doi.org/10.1093/biomet/73.2.461
55. Chan, K.S.: J. R. Stat. Soc. Ser. B Methodol. **53**(3), 691–696 (1991)
56. Wald, A., Wolfowitz, J.: Ann. Math. Stat. **11**(2), 147 (1940). https://doi.org/10.1214/aoms/1177731909
57. Schreiber, T., Schmitz, A.: Physica D **142**(3), 346 (2000). https://doi.org/10.1016/S0167-2789(00)00043-9
58. Zivot, E., Wang, J.: Modeling Financial Time Series with S-Plus®, vol. 191. Springer, Berlin (2007)
59. Trapletti, A., Hornik, K.: Tseries: Time Series Analysis and Computational Finance (2016). R package version 0.10-35
60. Friedman, M.: Ann. Math. Stat. **11**(1), 86 (1940). https://doi.org/10.1214/aoms/1177731944
61. Santos, T., Walk, S., Helic, D. In: Proceedings of the 26th International Conference on World Wide Web Companion, pp. 1567–1572 (2017). https://doi.org/10.1145/3041021.3051117
62. Christakis, N.A., Fowler, J.H.: N. Engl. J. Med. **358**(21), 2249 (2008). https://doi.org/10.1056/NEJMsa0706154
63. Aral, S., Walker, D.: Science **337**(6092), 337 (2012). https://doi.org/10.1126/science.1215842
64. Wagner, C., Mitter, S., Körner, C., Strohmaier, M.: Making Sense of Microposts (# MSM2012), p. 2 (2012)
65. Atkinson, K.E.: An Introduction to Numerical Analysis. Wiley, Hoboken (2008)
66. Galán-García, P., de la Puerta, J.G., Gómez, C.L., Santos, I., Bringas, P.G. In: International Joint Conference SOCO'13-CISIS'13-ICEUTE'13, pp. 419–428. Springer, Berlin (2014). https://doi.org/10.1007/978-3-319-01854-6_43
67. Seah, C.W., Chieu, H.L., Chai, K.M.A., Teow, L.N., Yeong, L.W. In: 18th International Conference on Information Fusion (Fusion), 2015, pp. 792–799. IEEE, Piscataway (2015)
68. Cambria, E., Chandra, P., Sharma, A., Hussain, A.: Do not feel the trolls. In: Proceedings of the 3rd International Workshop on Social Data on the Web (SDoW2010). CEUR Workshop Proceedings, vol. 664. CEUR-WS (2010)
69. Koncar, P., Walk, S., Helic, D., Strohmaier, M. In: Proceedings of the 26th International Conference on World Wide Web, WWW 2017, Perth, Australia, April 3–7, 2017, Companion Volume (2017). https://doi.org/10.1145/3041021.3051116
70. Laurent, G., Saramäki, J., Karsai, M.: From calls to communities: a model for time-varying social networks. Eur. Phys. J. B **88**(11), 301 (2015). https://doi.org/10.1140/epjb/e2015-60481-x
71. Moinet, A., Starnini, M., Pastor-Satorras, R.: Burstiness and aging in social temporal networks. Phys. Rev. Lett. **114**(10), 108701 (2015). https://doi.org/10.1103/PhysRevLett.114.108701
72. Karsai, M., Perra, N., Vespignani, A.: Time varying networks and the weakness of strong ties. Sci. Rep. **4**, 4001 (2014). https://doi.org/10.1038/srep04001
73. Bianconi, G., Darst, R.K., Iacovacci, J., Fortunato, S.: Triadic closure as a basic generating mechanism of communities in complex networks. Phys. Rev. E **90**(4), 042806 (2014). https://doi.org/10.1103/PhysRevE.90.042806
74. Kumpula, J.M., Onnela, J.P., Saramäki, J., Kaski, K., Kertész, J.: Emergence of communities in weighted networks. Phys. Rev. Lett. **99**, 228701 (2007). https://doi.org/10.1103/PhysRevLett.99.228701
75. Wölbitsch, M., Walk, S., Helic, D. In: International Workshop on Complex Networks and Their Applications, pp. 53–364. Springer, Cham (2017). https://doi.org/10.1007/978-3-319-72150-7_29

Interaction Prediction Problems in Link Streams

Thibaud Arnoux, Lionel Tabourier, and Matthieu Latapy

Abstract The problems of link prediction and recovery have been the focus of much work during the last 10 years. This is due to the fact that these questions have a large number of practical implications ranging from detecting spam emails, to predicting which item is selected by which user in a recommendation system. However, considering the highly dynamical aspect of complex networks, there is a rising interest not only for knowing who will interact with whom, but also when. For example, when trying to control the spreading of a virus in a population, it is important to know whether an individual is bound to have a lot of new contacts before or after being infected. In that sense, this question is located at the crossroad of link prediction and another family of problems which has been widely dealt with in the literature, that is, time-series prediction. We name it the interaction prediction problem in link streams. It calls for the definition of specific features, strategies, and evaluation methods to capture both the structural and temporal aspects of the interactions. In this chapter, we propose a general formulation of the problem, consistent with the link stream formalism, which formally represents the streaming sequence of interactions between the elements of the system. Using this framework, we discuss the formulation of the interaction prediction problem and propose possible strategies to address it.

Keywords Link stream · Interaction prediction · Link prediction · Time-series prediction

1 Introduction

Analyzing interactions over time plays a key role in many contexts: recommender systems (who buys which product and when), contacts between individuals

T. Arnoux · L. Tabourier (✉) · M. Latapy
Laboratoire d'Informatique de Paris 6, Sorbonne Université, CNRS, UMR7606, Paris, France
e-mail: thibaud.arnoux@lip6.fr; lionel.tabourier@lip6.fr; matthieu.latapy@lip6.fr

© Springer Nature Switzerland AG 2019   135
F. Ghanbarnejad et al. (eds.), *Dynamics On and Of Complex Networks III*,
Springer Proceedings in Complexity, https://doi.org/10.1007/978-3-030-14683-2_6

(message exchanges, physical proximity, or phone calls, for instance), and transaction analysis (like money or data transfers) are typical examples. As a consequence, much effort is devoted to the analysis of such data with approaches like temporal networks, time-varying graphs, or link streams [2, 4, 6].

Predicting future interactions is a crucial question in all these contexts, but the problem is traditionally addressed by merging interactions into a graph or series of graphs, called snapshots [7, 9, 12]. This has the key advantage of building a bridge with the powerful formalism and tools of graph theory, but at the cost of important information losses. More importantly, we argue that this approach misses interesting variants of the problem itself.

The goal of this chapter is to deepen our understanding of these interaction prediction problems. To do so, we formalize them within the link stream framework, which makes it possible to fully capture both the temporal and structural nature of data. This leads to several meaningful problem definitions that raise quite different challenges, as well as relations between them and classical approaches.

We focus here on problem definitions and comparisons; resolving some of them has already received attention [1, 3, 5] but unifying them into the same framework leads to a better understanding of the whole and the identification of new variants of interest. We also show that this helps to identify general approaches to tackle them.

Throughout this chapter, we assume a standard approach for solving prediction problems. First, one designs a model in order to make a prediction based on the fundamental assumption that future behaviors can be predicted from past observations. Then, parameters of the model are learned from past data, using an optimization process which aims at maximizing the prediction quality. Therefore, each prediction problem demands several ingredients, among which a quality estimator, features to describe past data, and a model to combine these features.

We first introduce the data modeling with link streams, which is the framework that we choose to address the interaction prediction problems. Afterward we present the prediction problems themselves, classified with respect to their ambition in the prediction task and we also discuss the subtle question of prediction evaluation. Finally, we propose a general direction for solving these problems using what we call *pairwise likeliness functions*.

2   Link Stream Modeling of Interactions

We use here the instantaneous link stream formalism presented in [6], which is a special case of stream graphs where nodes are always present and links have no duration. Such a link stream L is defined as a triplet (T, V, E), where $T = [\alpha, \omega] \subseteq \mathbb{R}$ is a time interval, V is the set of nodes under concern, and $E \subseteq T \times V \otimes V$ is a set of links: $(t, uv) \in E$ means that u and v interacted at time t. We consider here undirected interactions between pairs of distinct nodes u and v, which we denote by $uv \in V \otimes V$. We assume that E is finite: it contains a finite number of interactions,

Fig. 1 An example of undirected instantaneous link stream like the ones considered in this chapter: $L = (T, V, E)$ with $T = [0, 8]$, $V = \{a, b, c, d\}$, and $E = \{(0, ab), (1, bd), (2, ac), (3, bc), (5, ac), (5, cd), (7, bd), (8, ab)\}$

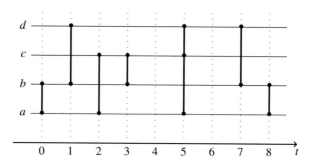

each occurring between two distinct nodes at a specific time instant. We illustrate this modeling in Fig. 1. Extending our work to more general cases is left for future work.

Such a link stream $L = (T, V, E)$ naturally induces a graph $G = (V, E')$ defined by $E' = \{uv : \exists t, (t, uv) \in E\}$: it is the graph in which two nodes of V are linked together if they interacted at some time in T. Dually, the link stream induces a time series $(\ell_t)_{t \in T}$ defined by $\ell_t = |\{uv : (t, uv) \in E\}|$: ℓ_t is the number of interactions occurring at time t.

In this context, the classical link prediction problem in graphs consists in predicting from G the links that will appear in the future, and the classical time-series prediction problem consists in predicting from $(\ell_t)_{t \in T}$ the number of interactions that will appear in the future. Our aim here is to draw benefit from L to predict richer information on future interactions. Depending on the targeted information, this leads to several, quite different problems that we detail in the next section.

3 Prediction Problems and Evaluation

Throughout the rest of this chapter, we assume that the set of nodes V remains unchanged, in other words nodes do not appear nor disappear. All the prediction problems that we consider start with an *input stream* $L_i = (T_i, V, E_i)$ with $T_i = [\alpha_i, \omega_i]$ and $E_i \subseteq T_i \times V \otimes V$. A prediction is related to an *output stream* $L_o = (T_o, V, E_o)$ with $T_o = [\alpha_o, \omega_o]$ and $E_o \subseteq T_o \times V \otimes V$. The time interval T_o is called the *prediction period*. Interactions actually occur during this period; we model them as the *actual stream* $L_a = (T_a, V, E_a)$ with $T_a = T_o$ and $E_a \subseteq T_a \times V \otimes V$. In addition, we always assume here that $\omega_i \leq \alpha_o$; in other words, we focus on predicting *future* interactions. Within this framework, the prediction is considered as good if the properties of L_o are similar to those of L_a.

3.1 Predicting All Interactions

3.1.1 Description

Predicting all interactions of all pairs in the actual stream may be the most ambitious formulation of the problem. It means that we aim at predicting each appearing link, i.e., predicting the stream L_a. We represent in Fig. 2 the situation corresponding to a given prediction.

3.1.2 Quality Evaluation

To evaluate the quality of such a prediction, a measurement of the distance between L_a and L_o is necessary. For a given pair of nodes, the series of actual or predicted interactions between them comes down to a set of points in $T_a = T_o$. Consequently, one may use a distance between two sets of points to evaluate the distance between the streams, and thus the prediction quality.

Among possible choices, let us mention the nearest point distance: the distance from a point $x \in X$ to the set Y is the distance from x to the closest point in Y, and the distance from X to Y is the sum of the distances of each $x \in X$ to Y. Though simple, this measurement is not formally a distance as it is not symmetric. Therefore,

Fig. 2 Top: schematic representation of the input stream L_i and the corresponding actual stream L_a. Bottom: schematic representation of the input stream L_i and the corresponding output stream L_o. The problem of predicting each interaction leads to comparing L_a to L_o

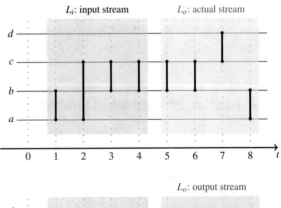
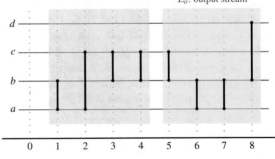

Fig. 3 Minimal cost
transformation of a set of
time points X into another Y:
the first step is deleting a
point from X (cost = fixed
penalty), and the other steps
consist in translating points
along the time axis (cost =
translation time distance)

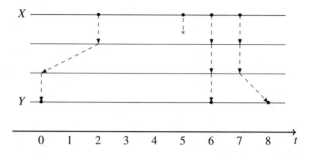

we may use instead the spike train distance proposed by Victor and Purpura [14],
which was originally designed to evaluate how different two neuronal impulse trains
are. We suggest to define the distance between link streams as

$$D(L_o, L_a) = \sum_{uv \in V \otimes V} D^{uv}(L_o, L_a),$$

where $D^{uv}(L_o, L_a)$ is the spike train distance between the points representing the
interactions between u and v in L_o and L_a. $D^{uv}(L_o, L_a)$ is defined as the minimal
cost to transform one set of points into the other with elementary steps: either
deleting, adding, or moving points along the time axis. Finally, $D(L_o, L_a)$ can be
understood as the minimal cost to transform L_o into L_a with these elementary steps.
When attributing a fixed cost to the addition and deletion steps, it is a metric distance
(see [14] for more details). To give the reader a more precise idea of the meaning of
this distance without diving in too much technical details, we represent in Fig. 3 an
example of minimal cost transformation of a set of time points into another.

3.2 Predicting the Next Interaction for Each Pair of Nodes

3.2.1 Description

A less constrained version of the former prediction task consists in predicting only
the next interaction for each pair of nodes (if it exists). Indeed, in many contexts an
experimenter is mostly interested in the moment when the next interaction occurs,
as represented in Fig. 4. For example, when predicting interactions in order to spread
an information through a network, the experimenter is interested in knowing when
the next interaction happens to spread the message as soon as possible. This task
has the advantage to circumvent the difficult prediction of the number of links per
pair of nodes.

In this case, the output of the prediction is not a stream, but the next occurrence
time for each pair $uv \in V \otimes V$. In order to include the case where there is no
interaction between u and v during the time interval of prediction T_o, a legitimate

Fig. 4 Top: schematic
representation of the input
stream L_i and the
corresponding actual set of
next occurrence times
$\{t_a^{uv}\}_{uv}$. Bottom: schematic
representation of the input
stream L_i and the
corresponding output set of
next occurrence times
$\{t_o^{uv}\}_{uv}$. The problem of
predicting the next interaction
for each pair of nodes leads to
comparing $\{t_a^{uv}\}_{uv}$ to $\{t_o^{uv}\}_{uv}$

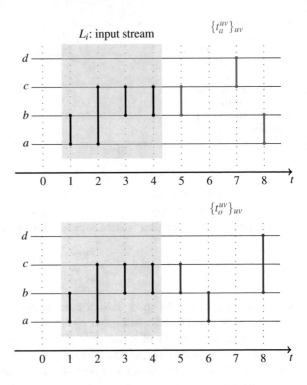

definition for the object predicted is the set $\{t_o^{uv}\}_{uv\in V\otimes V}$, with $t_o^{uv}\in[\alpha_o,\omega_o]\cup\{\infty\}$,
with $t_o^{uv}=\infty$ meaning that we predict no interaction for uv.

3.2.2 Quality Evaluation

In terms of quality evaluation, we should quantify the difference between the sets
$\mathscr{T}_o=\{t_o^{uv}\}_{uv\in V\otimes V}$ predicted and $\mathscr{T}_a=\{t_a^{uv}\}_{uv\in V\otimes V}$ actually occurring. Point set
distances such as the ones proposed in the previous task can be used here too, and
they are simpler with this prediction task, considering the fact that we take into
account at most one interaction for each link uv.

We denote d a distance function between two points in time. Then a possible
distance which can be seen as an equivalent of the spike train distance in this simpler
case is

$$D(\mathscr{T}_a,\mathscr{T}_o)=\sum_{uv\in V\otimes V}d(t_a^{uv},t_o^{uv})=\sum_{uv\in V\otimes V}min(|t_a^{uv}-t_o^{uv}|,p).$$

Using $d(t_a^{uv},t_o^{uv})=min(|t_a^{uv}-t_o^{uv}|,p)$ means that the distance between t_a^{uv} and t_o^{uv}
is either the delay between these interaction times or a predefined penalty p if there
is no interaction between u and v in L_a and we predicted one in L_o (and *vice versa*).

Fig. 5 Representation of the level of similarity between a predicted and an actual link, and its interpretation in terms of true and false positive prediction. In the situation represented, the interaction is predicted at instant t_o, which occurs before the actual interaction at t_a, and the prediction is thus both a true positive and a false positive to a certain extent. This extent is computed by a sigmoid-like similarity function

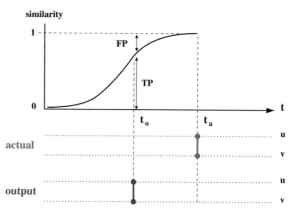

Thus it is similar to the addition/deletion cost of the spike train distance mentioned in Sect. 3.1. Here again, this quality evaluation is a simple and natural choice from our point of view, but other choices are available.

With this evaluation, the distance depends linearly on the time gap between a predicted link and a link observed in the actual stream. However, a user might consider that a linear dependence is not appropriate to describe the problem accurately and that other functions might be more relevant. In Fig. 5, we represent the case of a sigmoid-like distance function of the time gap. This distance function is complementary to a similarity function $s(x, y)$ such that $d(x, y) = 1 - s(x, y)$ with $y < x$, which is represented on the figure.

According to the evaluation method described above, we interpret the quality of the prediction using a notion of temporal distance between two events. Another possible interpretation of this evaluation method consists in using the vocabulary of classification tasks, as what is done in the case of link prediction problems. Indeed, if a link uv is observed in the actual stream at instant t_a^{uv} while it is not predicted yet, it can be interpreted as an equivalent of a *false negative*. Conversely, a link which is predicted while it is not observed yet is a *false positive*.

Of course, a link almost never occurs at the exact time when it has been predicted. Consequently, it is desirable not to use a 0/1 notion of false positive or false negative, but rather a score in the interval $[0, 1]$ which accounts for how close or how far we are from an exact prediction. That is what the similarity function s defined above does. To describe in more details the parallel between the classification-based to the distance-based interpretations, $s(t_a^{uv}, t_o^{uv})$ with $t_a^{uv} > t_o^{uv}$ quantifies the similarity, i.e., the degree of correctness of the prediction, while $1 - s(t_a^{uv}, t_o^{uv})$ represents the degree of error as a *false positive FP* does. If $t_o^{uv} > t_a^{uv}$, $1 - s(t_o^{uv}, t_a^{uv})$ rather represents the degree of error as a *false negative FN* prediction does. The degree of correctness can be mapped to the notion of *true positive TP*, which is consistent with the fact that $s(t_a^{uv}, t_o^{uv}) = 1$ when the link has been predicted exactly at the right time. Using this framework of interpretation, an unpredicted link is equivalent to a link predicted at $t_o^{uv} = \infty$, and similarly a non-occurring link is equivalent to a link occurring at $t_a^{uv} = \infty$.

It should be noted that *TP*, *FP*, and *FN* are usually boolean values which are defined unambiguously, while here the result depends on the choice of the distance function d. Besides that, *true negative (TN)* predictions do not have any obvious equivalent using temporal distances. However, it is enlightening to interpret the prediction with both the vocabulary of classification and temporal distances.

3.3   Predicting the Number of Interactions for Each Pair of Nodes

3.3.1   Description

Rather than predicting *if and when* each pair interacts, another relevant task consists in predicting *how many times* each pair interacts in a given period. It is less ambitious than the previous tasks, in the sense that we do not request to predict the exact time of links occurrence. In this case, the temporal precision of the prediction only depends on the duration of the output stream, and as it can be adjusted in the prediction protocol, we can tune how precise the prediction is in regards to the temporal dimension.

To formalize more precisely the prediction task, we define the notion of *activity of a pair of nodes uv* in the stream $L = (T, V, E)$ as $\mathscr{A}^{uv} = |\{(t, uv) \in E\}|$. In this context, our goal is that for any $uv \in V \otimes V$, $\mathscr{A}_o^{uv} = \mathscr{A}_a^{uv}$. Note that the activity quantifies the multiplicity of interactions between two nodes, so it is often represented by the weight of a link in the graph formalism.

3.3.2   Quality Evaluation and Relation to the Link Prediction Problem

In this case, distance measures such as the spike train distance cannot be used directly, as we do not predict interaction times. This task is actually closer to a more usual link prediction task on a graph snapshot, where the snapshot length corresponds to the duration of the actual stream, and we can draw advantage from that. We design quality estimators in the same spirit as what has been done in Sect. 3.2.2, by defining equivalents to *TP*, *FN*, or *FP* predictions.

As *FP* correspond to events that do not happen but are predicted, it is legitimate to translate this idea as the difference between the number of predicted links and the number of actual links if the former is larger than the latter. Similarly, *FN* correspond to events which occur but are not predicted, so it translates to the opposite difference if there are more actual links than there are predicted links. *TP* are the events which occur and are predicted so it is equivalent to the minimum between these two activities. Formally:

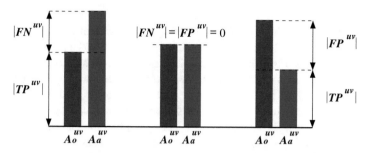

Fig. 6 Illustrations of the three possible cases ($\mathscr{A}_a^{uv} > \mathscr{A}_o^{uv}$, $\mathscr{A}_a^{uv} = \mathscr{A}_o^{uv}$, and $\mathscr{A}_a^{uv} < \mathscr{A}_o^{uv}$) of the interpretations of *TP*, *FN*, or *FP* in the context of the prediction of the number of interactions for each pair of nodes

$$\begin{cases} |TP^{uv}| = \min(\mathscr{A}_a^{uv}, \mathscr{A}_o^{uv}) \\ |FP^{uv}| = \max(\mathscr{A}_o^{uv} - \mathscr{A}_a^{uv}, 0) \\ |FN^{uv}| = \max(\mathscr{A}_a^{uv} - \mathscr{A}_o^{uv}, 0). \end{cases}$$

These definitions are illustrated in Fig. 6.

The definitions of *true positive*, *false positive*, and *false negative* proposed here satisfy usual relationships concerning these indicators: $|TP^{uv}| + |FP^{uv}|$ is the number of predicted interactions, and $|TP^{uv}| + |FN^{uv}|$ is the number of interactions between u and v in the actual stream.

Then, we denote $|TP|$ (resp. $|FN|$, $|FP|$) the total number of true positive (resp. false negative, false positive) in the stream:

$$\begin{cases} |TP| = \sum_{uv \in V \otimes V} |TP^{uv}| \\ |FN| = \sum_{uv \in V \otimes V} |FN^{uv}| \\ |FP| = \sum_{uv \in V \otimes V} |FP^{uv}|. \end{cases}$$

We can thus define accordingly useful quantities to evaluate the quality of a prediction:

- precision: $\frac{|TP|}{|TP|+|FP|}$, which represents the fraction of good predictions among the total number of predictions,
- recall: $\frac{|TP|}{|TP|+|FN|}$, which represents the fraction of events detected among the total number of events which can be detected, and
- F1-score, which is the harmonic mean of the precision and recall, that is, to say $F_1 = 2 \cdot \frac{precision \cdot recall}{precision + recall}$.

Using this interpretation, a good prediction can be considered, for example, as a prediction that maximizes the F1-score, as it reflects a compromise between precision and recall. Nevertheless, we do not define any equivalent to a *true negative* prediction and to the total number of negative predictions in general. It makes us unable to define equivalents of other classification estimators (fall-out, specificity, ROC curve, etc.).

As stated previously, this problem relates to the weighted link prediction problem: given a weighted graph representing the number of interactions between each pair of nodes, predict the future weight. Related problems exist in the link prediction literature. For instance, some authors have proposed to divide links into two families: new links and recurring links, and then make two separate predictions for each family [13]. Besides that, our task can also be related to the matrix completion problem, which is usually considered with boolean adjacency matrices in the context of link prediction (see, for example, [10]) but can be generalized to matrices with positive values. Powerful as they may be, these approaches leave in the shadow the fundamentally temporal nature of the data, which our formulation of the problem tries to grasp.

3.4 Predicting the Existence of Interaction(s) for Each Pair of Nodes

Finally, a natural problem is predicting if a pair interacts at least once in the actual stream. Another way of formulating this task using the activity defined in Sect. 3.3 could be to predict for all pairs of nodes if they reach an activity of 1 during the prediction period. An interesting point concerning this prediction task is that it is actually similar to the classical link prediction problem: the prediction task comes to predicting the structure of the actual graph $G_a = (V, E'_a)$ aggregated from L_a, where uv is in E'_a if there is at least one (t, uv) in L_a. The main difference is that the link stream formalism stresses the fact that both structural and temporal information can be used as features to improve the prediction quality. Such information has already been used in the literature in order to achieve link prediction tasks, but in a more classical framework (see, e.g., [8, 9]).

In terms of evaluation, link prediction tasks have been widely studied as binary classification tasks, and thus one makes use of the evaluators usually employed for such issues (precision, recall, F-scores, ROC curve, etc.).

4   Pairwise Likeliness Functions for Prediction Tasks

From now on, we suppose that the prediction problem and its evaluation method are set, and we focus on the prediction model. We present in this section a possible way to address these prediction problems taking into account the fact that the data contains structural and temporal information. Consequently, the features of the stream that we use for prediction should be described in a way that can reflect both structural and temporal properties.

The overall approach is the following. First, we compute pairwise likeliness functions using properties of the input stream L_i. Pairwise likeliness functions are designed to reflect when we expect a pair of nodes to interact during the period T_o. The prediction model relies on these pairwise likeliness functions: one would train the parameters of the model by maximizing the quality of the prediction on a learning stream, L_a, using the vocabulary defined in Sect. 3. After this learning phase, the model can be used for prediction.

4.1 Pairwise Likeliness Functions

In order to represent a feature of the input stream on which the prediction is based, we use a function $f^{uv}(t)$ which represents the likeliness for a link uv to occur at time t. An interesting aspect of this approach is that it gathers in a same formalism both temporal and structural (and potentially hybrid) features. We call such functions *pairwise likeliness functions*.

4.1.1 Illustration

To set the reader's mind, we illustrate this notion using three examples. Let us consider the following features:

1. a structural feature often used in link prediction problems: the *number of common neighbors* shared by two nodes,
2. a temporal feature based on the assumption that there is some regularity in the temporal patterns of interaction between two nodes that we call *regularity*,
3. another temporal feature which is used to reflect the fact that there are episodes of bursty activity of interactions, the *burstiness*.

For these examples, we suggest possible definitions of the corresponding pairwise likeliness functions. These definitions are based on common sense, but other possibilities would make sense. Our goal here is to show that this formalism is versatile.

1. Concerning the number of common neighbors, the pairwise likeliness function is a constant (independent from time), which is simply the number of common neighbors itself

$$f_{CN}^{uv}(t) = |\{w : \exists\, (t_1, uw), (t_2, wv) \in E_i\}|$$

2. Regularity is defined using the interaction times between u and v during the input stream. Supposing that the links are approximately regularly spaced, a consistent shape for the likeliness function could be a sinusoidal function, as sketched in Fig. 7. The corresponding definition is

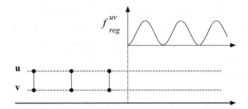

Fig. 7 Representation of a sinusoidal regularity-based likeliness function. Bottom: input stream. Top: corresponding regularity-based likeliness function computed from the input

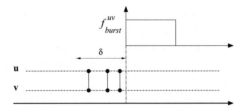

Fig. 8 Representation of a rectangular burstiness-based likeliness function. Bottom: input stream. Top: corresponding burstiness-based likeliness function computed from the input

$$f_{reg}^{uv}(t) = \frac{1}{2} + \frac{1}{2}cos\left(\frac{2\pi(t - t_\ell)}{\langle \tau \rangle}\right).$$

where t_ℓ denotes the last interaction time of uv in L_i and $\langle \tau \rangle$ the average interaction time during T_i.

3. Finally concerning burstiness, we consider that if a train of interactions (that is to say more than two interactions) has begun less than a time δ ago, then there is an increased probability of interaction during the next δ duration, as represented in Fig. 8. The corresponding definition is

$$f_{burst}^{uv}(t) = \begin{cases} 1 \text{ if } t \in [0; \delta] \text{ and } | \{(t, uv) \in E_i \text{ with } t \in [-\delta; 0]\} | > 2 \\ 0 \text{ else.} \end{cases}$$

4.1.2 Combining Pairwise Likeliness Functions

In order to achieve the prediction itself, we now define a prediction model based on pairwise likeliness functions. We combine these functions into $\mathscr{F}^{uv}(t)$, the *combined pairwise likeliness function*. Again, there are many possible ways to achieve this combination, we propose to use a linear combination as an illustration of the approach:

Fig. 9 Illustration of a
combined pairwise likeliness
function for a pair of nodes
uv, based on the linear
combination of f_{CN}, f_{reg},
and f_{burst}

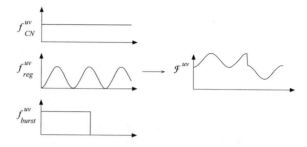

$$\mathscr{F}^{uv}(t) = a_{CN} \cdot f_{CN}^{uv}(t) + a_{reg} \cdot f_{reg}^{uv}(t) + a_{burst} \cdot f_{burst}^{uv}(t).$$

In this framework, the coefficients represent the weight given to the different features in the combination. On the examples of the three pairwise functions previously defined, the combination for one pair uv is represented graphically in Fig. 9.

4.2 Combined Pairwise Likeliness Functions for Prediction Tasks

Now that, for each pair of nodes, we have a function representing the likeliness of an interaction during the prediction period, we discuss how this function can be used to achieve the prediction tasks formerly presented.

We have seen in Sect. 3 that there are two different kinds of tasks: On the one hand, predicting the appearance of one or several links, that is, to say predicting precisely the triplets (t, uv) (tasks 1 and 2); on the other hand, predicting the number of links which occur during a given period of time (tasks 3 and 4). We discuss these two families of prediction tasks separately.

4.2.1 Predicting One or Several Link Occurrences

Given a pair of nodes, the goal is to predict what are the occurrence times—if any— of interactions between these nodes. A natural way could be to detect the local maximum of the combined likeliness function. In this case, the problem seems to map to detecting peaks in a function equivalent to a time series. A given point of a time series is said to be a peak if the associated value is larger than a specified threshold. Peak detection is an active area of research, and techniques could be derived from this field (see, for example, [11]). A problem is that such methods aim at identifying points which stand out from their neighbors in the time series, while here a user would rather consider that a long plateau should correspond to the existence of one or even several interactions. In other words, the problem is not exactly equivalent to the intuition of a peak detection method. It may be closer to the

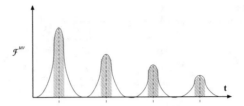

Fig. 10 Illustration of a criterion to select significant peaks: the areas colored are the area under $\mathscr{F}^{uv}$ centered on a local maximum and of width 2δ. Areas in blue are larger than a threshold Θ, while areas in brown are smaller than Θ

burst detection problem (e.g., [15]): one looks for a time window during which the aggregated signal is larger than a user-specified threshold. However, burst detection usually focuses on locating a period of high activity, rather than a precise point in time. Both tasks are thus not identical in that case too.

In any case, there should be additional criteria to consider if a peak is significant enough to justify the prediction of an interaction. One way of doing so is to define an area under the curve around the peak, and the peak is considered significant if this area is larger than a threshold, as schematically represented in Fig. 10. Formally, if a peak has been detected at time τ, the criterion of significance would be $\int_{\tau-\delta}^{\tau+\delta} \mathscr{F}^{uv}(t)\,dt \geq \Theta$, where Θ is the significance threshold of the area around the peak, and δ characterizes the width defining the area under the curve around the peak. The choices of δ and Θ are obviously critical for the prediction task, as it will largely influence the number of links in the predicted stream. The parameters of any method should thus be carefully chosen based on the input stream.

Note that predicting only the next interaction does not alleviate the problem of peak significance mentioned above and also calls for non-trivial choices to decide if a peak is significant enough to justify the existence of an interaction.

Another issue should be mentioned when several link occurrences are predicted. Interactions are not independent from each other, meaning that if an interaction is predicted at time t, it should affect the probability for an interaction to occur at any time $t' > t$. This issue is simply ignored when the problem is managed as a peak detection problem, which is another limitation to this technique. To address the issue, it is possible to predict interactions sequentially, first considering the next occurrence, then assuming it does happen in order to predict the next one, etc. However, other difficulties appear when tackling the problem along these lines, one of which being the accumulation of prediction errors throughout the process.

4.2.2 Predicting the Number of Interactions Over a Given Period

When considering the prediction of a number of links during a given period, one would certainly use the likeliness functions differently. As we no longer predict the interaction occurrence times, it is not necessary to detect the peaks of the likeliness function.

The area under the likeliness function curve represents the likeliness for a link to appear over the whole prediction period. Therefore, one could consider that this area should be related to the number of interactions actually occurring during that period. So, supposing that we are able to predict efficiently the total number of interactions in the prediction stream, it is possible to predict the number of links for any pair uv by allocating links to pairs of nodes proportionally to the area under the likeliness function curve. In short, a relevant relation for predicting the activity of a pair uv over the prediction period $T_o = T_a = [\alpha_o, \omega_o]$ is to consider that

$$\mathscr{A}_o^{uv} = C \cdot \int_{\alpha_o}^{\omega_o} \mathscr{F}^{uv}(t)\,dt,$$

where C is a constant fixed by the total number of interactions in the output stream. As predicting this number is a classical time-series prediction task, we have in our hands the tools to achieve the prediction of the number of links for any pair of nodes in the stream during a given period.

5 Conclusion

In this chapter, we have formulated the problem of predicting interactions in a link stream, which can be seen as a generalization of the link prediction problem in a network when the temporal dimension of the data is taken into account, or dually, as the generalization of a time-series prediction, when there is a network-like structure supporting the various time series.

We have seen that the most general problems, predicting exactly the moments when two nodes in the stream will interact with each other, is certainly a difficult task to achieve—as could be expected. But we have also proposed different, more humble tasks, which seem simpler to address as they are closer to more classical prediction problems, namely, the link prediction task in a graph. Precisely, the task of predicting the number of links which appear during a given period of time seems promising. Indeed, it allows to use evaluation metrics which can be interpreted to a certain extent using the vocabulary of classification tasks, and we presented a possible way to tackle this prediction using features of the input stream that would account for its structural and temporal characteristics.

We do not develop in this chapter the details of the technical implementation of this method. However, an interested reader can refer [1] for a more comprehensive view of an implementation on contact networks, which suggests that there are indeed good prospects (and still a lot to do) on these prediction tasks.

References

1. Arnoux, T., Tabourier, L., Latapy, M.: Predicting interactions between individuals with structural and dynamical information (2018). Preprint. arXiv:1804.01465
2. Casteigts, A., Flocchini, P., Quattrociocchi, W., Santoro, N.: Time-varying graphs and dynamic networks. Int. J. Parallel Emergent Distrib. Syst. **27**(5), 387–408 (2012)
3. da Silva Soares, P.R., Cavalcante Prudêncio, R.B.: Time series based link prediction. In: The 2012 International Joint Conference on Neural Networks (IJCNN), pp. 1–7. IEEE, Piscataway (2012)
4. Holme, P., Saramäki, J.: Temporal networks. Phys. Rep. **519**(3), 97–125 (2012)
5. Huang, Z., Lin, D.K.J.: The time-series link prediction problem with applications in communication surveillance. INFORMS J. Comput. **21**(2), 286–303 (2009)
6. Latapy, M., Viard, T., Magnien, C.: Stream graphs and link streams for the modeling of interactions over time. (2017, preprint). arXiv:1710.04073
7. Liben-Nowell, D., Kleinberg, J.: The link-prediction problem for social networks. J. Assoc. Inf. Sci. Technol. **58**(7), 1019–1031 (2007)
8. Lichtenwalter, R.N., Lussier, J.T., Chawla, N.V.: New perspectives and methods in link prediction. In: Proceedings of the 16th ACM SIGKDD International Conference on Knowledge Discovery and Data Mining, pp. 243–252. ACM, New York (2010)
9. Lü, L., Zhou, T.: Link prediction in complex networks: a survey. Phys. A Stat. Mech. Appl. **390**(6), 1150–1170 (2011)
10. Nie, F., Wang, H., Cai, X., Huang, H., Ding, C.: Robust matrix completion via joint Schatten p-norm and lp-norm minimization. In: 2012 IEEE 12th International Conference on Data Mining (ICDM), pp. 566–574. IEEE, Piscataway (2012)
11. Palshikar, G., et al.: Simple algorithms for peak detection in time-series. In: Proceedings of 1st International Conference on Advanced Data Analysis, Business Analytics and Intelligence, pp. 1–13 (2009)
12. Sarkar, P., Chakrabarti, D., Jordan, M.: Nonparametric link prediction in dynamic networks. (2012, preprint). arXiv:1206.6394
13. Scholz, C., Atzmueller, M., Stumme, G.: On the predictability of human contacts: influence factors and the strength of stronger ties. In; 2012 International Conference on and 2012 International Conference on Social Computing (SocialCom) Privacy, Security, Risk and Trust (PASSAT), pp. 312–321. IEEE, Piscataway (2012)
14. Victor, J.D., Purpura, K.P.: Metric-space analysis of spike trains: theory, algorithms and application. Netw. Comput. Neural Syst. **8**(2), 127–164 (1997)
15. Zhu, Y., Shasha, D.: Efficient elastic burst detection in data streams. In: Proceedings of the Ninth ACM SIGKDD International Conference on Knowledge Discovery and Data Mining, pp. 336–345. ACM, New York (2003)

The Network Source Location Problem in the Context of Foodborne Disease Outbreaks

Abigail L. Horn and Hanno Friedrich

Abstract In today's globally interconnected food system, outbreaks of foodborne disease can spread widely and cause considerable impact on public health. Food distribution is a complex system that can be seen as a network of trade flows connecting supply chain actors. Identifying the source of an outbreak of foodborne disease distributed across this network can be solved by considering this network structure and the dimensions of information it contains. The literature on the network source identification problem has grown widely in recent years covering problems in many different contexts, from contagious disease infecting a human population, to computer viruses spreading through the Internet, to rumors or trends diffusing through a social network. Much of this work has focused on studying this problem in analytically tractable frameworks, designing approaches to work on trees and extending to general network structures in an *ad hoc* manner. These simplified frameworks lack many features of real-world networks and problem contexts that can dramatically impact transmission dynamics, and therefore, backwards inference of the transmission process. Moreover, the features that distinguish foodborne disease in the context of source identification have not previously been studied or identified. In this article we identify these features, then provide a review of existing work on the network source identification problem, categorizing approaches according to these features. We conclude that much of the existing work cannot be implemented in the foodborne disease problem because it makes assumptions about the transmission process that are unrealistic in the context of food supply networks—that is, identifying the source of an epidemic *contagion* whereas foodborne contamination spreads through a transport network-mediated *diffusion* process, or because it requires data that is not available—complete observations of the contamination status of all nodes in the network.

A. L. Horn (✉)
Division of Epidemiology, Zoonoses and Antibiotic Resistance, Federal Institute for Risk Assessment (BfR), Berlin, Germany
e-mail: abbylhorn@alum.mit.edu

H. Friedrich
Kühne Logistics University, Hamburg, Germany

© Springer Nature Switzerland AG 2019
F. Ghanbarnejad et al. (eds.), *Dynamics On and Of Complex Networks III*,
Springer Proceedings in Complexity, https://doi.org/10.1007/978-3-030-14683-2_7

Keywords Complex systems · Network source identification · Epidemic ·
Disease spreading · Network diffusion · Food supply networks · Foodborne
disease

1 Introduction

An important problem for many networked systems involving spreading processes
is identifying the source of the spreading agent; if the contaminated food source,
patient zero, or the rumor originator is identified efficiently, damage can be
prevented or reduced [7, 10, 13, 25].

Over the past couple decades there has been significant effort devoted to
studying the dynamics of outbreaks on networks [5, 17, 19, 23, 24, 26, 31]; for a
comprehensive review of epidemic spreading on complex networks, see [27]; for
a review of information diffusion on complex networks including a comparative
evaluation of available models and algorithms, see [36]. Most of this work has
focused on the forward problem of understanding and forecasting the diffusion
process and its dependence on the structure of the underlying network. However
in recent years much work has emerged on the inverse problem of identifying the
source of an outbreak spread in a network. This work covers problems in different
contexts, including contagious disease infecting a human population; rumors or
information diffusing through a social network; adoption of an idea, behavior
change, or product in an organizational network; the spread of viruses on the
internet; and the transport-mediated diffusion of contaminated individuals between
cities. These contexts represent different spreading scenarios that require different
modeling approaches for forward dynamics and inverse solutions.

Most studies of spreading processes in networks have been done in the context of
epidemiology, modeling the spread of diseases or viruses through a host population.
Network disease propagation models are based on the stages of disease as it infects
individuals and spreads across contact links in a host population. Initially the
entire population is susceptible to the disease; once any individual is exposed to
an infectious contact they become infected and can infect others; from this point
they can recover, be removed, become immune, or other variants. These models
are referred to as compartmental models due to the disease compartments that
individuals move between in illness progression: S—susceptible, I—infected, R—
recovered or removed, etc.

Compartmental disease spreading models represent a simple contagion process,
because only one direct contact with an infected neighbor is required for the
contagion to be transmitted. Along with disease, information spread through a
network has been shown to follow a simple contagion process. On the other hand,
behavior change has been shown to spread as a complex contagion that requires
multiple sources of exposure or reinforcement for the new behavior to be adopted.

A typical quantity that is studied in relation to network epidemic models is the epidemic threshold, or the set of conditions under which the disease will either proliferate or die out in the network. Unlike classical diseases or viruses spread through social contact networks, computer viruses have been shown to have an epidemic threshold of 0, meaning that the infectivity rate can be vanishingly small for the epidemic to happen. This is due to the scale-free structure of computer networks, which are extremely heterogeneous with a few nodes having an extremely high number of connections. The spread of computer viruses therefore diverges from classical diseases not due to the contagion model—both are simple contagion— but due to the heterogeneity of the network substrate over which computer viruses spread.

Another type of epidemic model is the metapopulation reaction–diffusion process, which in addition to contagion dynamics accounts for the role of movement or transport in diffusing a contamination in space. In this type of model, nodes represent subpopulations, such as cities, and links represent the movement of individuals between subpopulations. Individuals interact in each subpopulation according to assumptions of equal mixing or a local social network structure and disease spreads between these individuals according to a contagion model; this is the reaction process. The movement of individuals between subpopulations is the spatial diffusion process, often modeled over a network as a Markov transition process. Metapopulation models therefore depend both on the local social network structure at each node and on the spatial structure of the environment, transport infrastructures, traffic networks, and other movement patterns over which individuals diffuse.

Approaches to the source detection problem are developed in the context of one of these forward spreading processes. Most approaches have been devised in the context of simple contagion processes including infectious disease outbreaks in human contact networks or rumors spreading in social networks [1–3, 20, 33, 37, 38]. Another stream of work has focused on identifying the source of processes in which network-mediated spatial diffusion is the main vector of spread. This includes contagious diseases spread through drift in water systems [28] or spreading between cities by global air travel [6], and foodborne disease contamination spread through food distribution networks [14].

This article focuses on foodborne disease. The features that distinguish foodborne disease in the context of source identification have not previously been studied or identified. In this work we identify these features and conclude that most of the existing approaches to source detection cannot be implemented in the foodborne disease problem because they make assumptions about the transmission process that are unrealistic in the context of food supply networks—that is, identifying the source of an epidemic contagion [1–3, 20, 33, 37, 38] whereas foodborne contamination spreads through a transport network-mediated network diffusion process, or because it requires data that is not available—complete observations of the contamination status of all nodes in the network [8, 11, 30, 34] or timed network data [1–3, 15, 20, 21, 28, 33, 35]. We begin by first providing relevant background on outbreaks of foodborne disease and the contamination diffusion process.

1.1 Large-Scale Outbreaks of Foodborne Disease

The complexity and globalization of food production have made foodborne disease a widespread public health problem worldwide. A small but worrisome minority of outbreaks are generated by a contamination originating at the site of production or processing, generating a widespread diffusion of contamination through the supply chain and affecting a potentially great number of people across geographically distributed locations. As recent trends continue, including large-scale production practices and distribution over ever-larger distances, both the frequency and the severity of consequences of large-scale outbreaks are increasing. In the USA, the number of large-scale (i.e., multi-state) outbreaks increased by 135% in the years 1995–2004 to the years 2005–2014. These large-scale outbreaks accounted for 3% of total outbreaks, which includes localized, non-distributed incidents, but were responsible for 34% of hospitalizations and 56% of deaths [9].

During a large-scale outbreak of foodborne disease, rapidly identifying the source, including both the food vector carrying the contamination and the location source in the supply chain, is essential to minimizing impact on public health and industry. However, tracing an outbreak to its origin is a challenging problem due to the complexity of the food supply system. Furthermore, current investigation methods represent a missed opportunity to utilize valuable information to solve the source localization problem.

Food distribution is a complex system that can be seen as a network of trade flows connecting supply network actors. Identifying the source of an outbreak of contamination distributed across a network can best be solved by considering this network structure and the dimensions of information it contains. Together with reports of illness, this network information can be used to solve the problem of identifying the source of large-scale outbreaks.

To formulate the problem of source detection on a network, assumptions must be made regarding (1) the network and observation data available for source identification, and (2) the transmission process that led to the observations. Based on basic practical knowledge of food supply networks and the foodborne disease contamination process, in this article we introduce the source identification problem in the context of foodborne disease outbreaks and outline six features that distinguish this problem from source detection in other network contexts due to either practical data limitations or differences in transmission process mechanics. We then use the six features to categorize the existing literature on the network-based source detection problem according to relevance to the foodborne disease context.

2   Background and Definitions

2.1   Network-Based Source Identification

To solve the source detection problem in the context of foodborne disease, a network model of the supply of a specific food commodity is assumed as a given input. A probabilistic model of the transmission process of contamination spreading through this network is then postulated. In the following, we assume that a foodborne disease outbreak will originate from a single contamination source. This source sends out contaminated products that travel through the network according to the transmission model, resulting in observations of illness at a set of network nodes. The source identification objective is to minimize the error between the model-derived estimate of the location of the source and the true location of the source in the network, given the nodes associated with the observations of illness.

2.2   Food Supply Networks and Foodborne Disease Transmission

Food supply systems can be represented by a directed network structure consisting of multiple stages of production, distribution, storage, and consumption. Flows through the network are generally structured such that product is distributed in a forward direction along a *path*, or a collection of directed *edges* connecting supply *nodes* from origination to point of sale. A large-scale outbreak occurs when contaminated food departs from some source in an early stage of the network that is able to reach downstream nodes in geographically distributed locations. The contamination will eventually make its way to consumers, who develop illness some time after consuming the contaminated food. Case reports of illness are associated with the supply network node at which the offending product was purchased and *exits* the supply network, e.g. a retailer or restaurant; these nodes can be considered *infected*.

The network in Fig. 1 represents a supply network in which contamination at a food producer has spread through the supply network, leading to reports of illness at three different retailers. With this structure mapped, it is straightforward to utilize all case data (i.e., evidence) available during an event to identify the set of *feasible* sources of contamination, that is, the set of nodes that connect to all known contaminated nodes. Network structural information thus provides a first cut into the source identification problem by enabling the identification of feasible sources. To differentiate between the feasible sources, further dimensions of information available within the network can be leveraged. Each edge contains information about the volume of goods traded between supply network actors. Volume-weighted information is a source of heterogeneity that can be thought of as the relative propagation potential of a given edge, providing insight into the paths along which contaminated product is likely to have traveled.

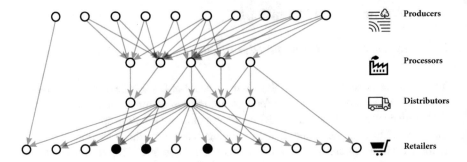

Fig. 1 Illustration of a food distribution network with three reported cases of illness (at the shaded nodes) linked to retailer nodes. Figure source: [8]

3 Distinguishing Features of Foodborne Disease Transmission

3.1 A Transport, Not Epidemiological, Transmission Process

Many network-based source detection methods are designed to identify the source of an infectious contagion. These methods often assume some variant of the epidemiological model of contagion transmission, including the widely used susceptible-infected (SI) or susceptible-infected-recovered (SIR) models. However the transmission of contamination through the food supply to people is different from the disease contagion process from people to people. Contamination spreads as contaminated (solid, perishable) food moves through the supply network after being inoculated by the pathogen at the source. As the food is transported through the supply network, the pathogenic quantity will generally remain conserved, meaning it will neither spread to other food items nor decay significantly in infectivity [18, 29]. The former is due to a number of factors including the lack of contact between packaged items, the lack of interaction or mixing between unpackaged items, and the biological insusceptibility of contamination to transmission and decay, i.e. low infectivity and recovery rates.

Due to this conservation of contamination, the spreading process in the context of foodborne disease primarily involves the contaminated food being spatially distributed along the network without decaying (i.e., recovery) or growing (i.e., infection) the contamination along the way. Contaminated food items cause infection in people when the food is consumed, but this process does not represent a classical infection dynamics because the contamination is directional (food to human) and largely does not spread between people. Contagion processes represent a different dynamics; if these are applied to the foodborne disease situation, the extremely low infection rate would mean that when individual food items come into contact, the infection will not be transmitted and will die out. The diffusion along the network is the mechanism that moves the contamination forward through the supply chain. To reflect these diffusion dynamics, the foodborne disease contamination spreading process has therefore been modeled as a simple Markov transmission process [14].

It would be possible to model the foodborne disease spreading process using a metapopulation reaction–diffusion process, as discussed in Sect. 1, where nodes represent locations in the supply chain containing a constant "subpopulation" of food items, and links represent the transport of food items between supply chain locations. However because contagious transmission is largely not occurring between food items, a metapopulation model would add more complexity (by incorporating the inactivated local contagion process along with the diffusion process) without incorporating more of the dynamics of the spread of contamination by food through the supply chain. Therefore in the following, we will refer to the foodborne disease contamination process as a diffusion-type process by which we mean exclusively network-mediated diffusion and not contagion.

Finally, the observation data available for source identification occurs on the human level and not on the food item level, and only via infection status, (I) in the SI/R model. Observations of contamination occur when people report illness. Each illness is linked to a supply network node at which the contaminated food was purchased. Data regarding the contamination status of individual food items is not normally available during an investigation. Furthermore, it is not possible to establish from the illness reports whether a supply node has ever received contaminated food and is thus susceptible (S), as it may have led to illnesses that went unreported. Methods that rely on observations of susceptible status or that assume nodes not reporting infection are contamination-free (also called "negative information") are thus non-applicable in this setting.

3.2 Observations are Sparse

Though the contamination will travel through multiple network nodes on its journey through the supply network, it is only observed when illness is reported in connection with the exiting or *absorbing* node at which contaminated food was purchased. The contamination status of *transient* nodes involved in the production, processing, or storage of food, though closer to the source in number of network edges, will remain hidden to investigators unless further investigations are performed (normally during later stages of an investigation). Furthermore, even at the consumption level, the overwhelming majority of foodborne illness cases are either not identified or logged by authorities, with official estimates of underreporting varying from 10 to 75 times for different pathogens [32]. A trivial implication of the sparsity of observations is that it is unrealistic to assume, as some source detection methods do, that the contamination status of all nodes in the network is known.

3.3   Observations will Always be Spaced Far from the Source

The placement of observations only at absorbing nodes also means that there will
be a large network distance between the source and each observation, increasing the
number of possible paths that could have been traveled and in turn the uncertainty
in the *structure* of the diffusion trajectory. At the same time, the differing volume-
weights along the edges of the supply network provide valuable information for
inference. Given the large uncertainty in the diffusion structure, approaches to
source detection that consider network structure alone will be inferior to those that
consider this weighted information.

3.4   Similar Path Lengths

Due to the staged structure of the food supply network, paths through the network
from source to observation will be close to the same length in terms of number of
network edges. This is common for supply chain networks of all types, and can
be observed in models of food supply networks across all product groups [4, 12].
Many existing source detection methods simplify the inference process by assuming
that the contamination traveled across the shortest path from the source to each
observation, or otherwise by leveraging shortest path properties of graphs. These
approximations will apply poorly in the food supply network context where most
paths will be indistinguishable in length.

3.5   Multiple Candidate Paths

Between any possible source and observation in a food supply network, there exist
multiple paths of travel of similar weight or likelihood. This is due to the lack of
monopolies in food production, trade, and retailing markets: any given food type
will be distributed through multiple larger retailers or wholesalers, each dealing
with similarly large volumes of product [4, 12]. Certain source detection methods
make the simplifying assumption that the contamination travels across the single
highest-probability path between a source and observation. These methods will be
inaccurate in the food supply network setting where transmission dynamics are not
necessarily dominated by a small percentage of connections.

3.6   Data on Times Through the Network are Lacking

In theory, there should be a signal for source detection from the timed reports of
illnesses combined with a model of the time it takes to transmit the contamination.
Each collection of edges in a network path encodes information about the time

delay that a contaminated product could have taken to travel these steps. These delays will be distributed differently according to parameters like the distance and speed of travel and supply network logistics encountered. However, there is significant temporal uncertainty in the contamination transmission process. The time the contamination may spend in storage, both at various nodes along the supply network (e.g., warehouses) and with the consumer after purchase, as well as during the incubation period, can be significant and vary widely—and potentially much more so than the time spent in travel. Furthermore, while the times of infection are available to some degree of accuracy (recorded according to patient recalled time of illness onset), data on storage times through the network are unavailable with the exception of a few case-specific customer or retailer survey studies [18, 29]. Therefore, while time can be an important aspect in some foodborne disease source detection applications, time-based methods are not currently implementable in the foodborne disease context given available data.

4   Categorization of Literature

Many approaches to the network source detection problem have been developed in recent years, though none of these methods have specifically considered the context of outbreaks of foodborne disease. We now review the major themes in the existing work, using the features described above to guide the discussion in terms of relevance to the problem on food supply networks. The categorization of existing work in terms of these features is summarized in Table 1.

The earliest approaches to source detection are based on complete observations, relying on knowing the contamination status (SI/R) of each node in the network at a fixed point in time [8, 11, 30, 34]. These methods do not incorporate information about differing weights along edges but are based solely on graph structure by employing notions of network centrality, the intuition being that the node most "central" to the observed contamination process is the source. The seminal work by Shah and Zaman [34] introduces the measure of *rumor centrality*, which considers the number of linear extensions between each source and the infected nodes. The method and analytical results concerning detection probability are derived for trees or tree-like graphs; to apply to general networks, a Breadth-First-Search (BFS) heuristic that assumes the contamination traveled across the shortest paths to the observations must be used. Other methods based on *betweenness centrality* [8] and *eigenvector centrality* [11, 30] apply to general networks without employing a shortest path heuristic, although the calculation of betweenness is based on shortest path properties. These methods were important for establishing foundational results on the network source detection problem but are impractical for real network-outbreak scenarios due to the complete observation assumption.

Many methods have since been developed for the more realistic setting that only a subset of the contaminated nodes are observable, i.e. partial observations. These can be categorized into temporal methods—approaches designed to make use of

Table 1 Categorization of existing work on the source detection problem according to relevance to the foodborne disease context

Source identification methodology of existing work	Limitations of source identification methodologies in foodborne disease context					
	(1) Only SI/R (no diffusion component)	(2) Assumes complete observations	(3) Ignores weights	(4) Only shortest paths	(5) Only dominant paths	(6) Assumes times through network
Rumor centrality [34]		X	X	X		
Betweenness centrality [8]		X	X	X		
Eigenvector centrality [11, 30]		X	X			
Message passing [20]	X					X
Belief propagation [2]	X					X
Analytic combinatoric [3]	X					X
Gaussian [21, 28]				X		X
Four-metric [33]	X			X		X
Monte Carlo [1]	X			X		X
Analytical time-varying networks [15]				X		X
Time-reversal backward spreading [35]						X
Jordan centrality [37, 38]	X			X		
Effective distance [6, 22]					X	
Multiple-paths diffusion [14]						

the information from the timed reports of illness and times through the network, and non-temporal methods—approaches that rely only on the node location where contamination has been reported. The temporal category includes methods assuming discrete-time epidemic (SI/R) contagion models based on *dynamic message-passing* [20], *Bayesian belief propagation,* [2], *analytic combinatoric* approaches [3]. The *analytic-combinatoric* method [3] builds on the approach of [20] and [2] by removing the node-independence assumption of [20] and the tree-like contact network assumption that both [20] and [2] are predicated on to compute the exact source probability distribution for general contact network structures. Because the analytical calculations increase exponentially for non-tree-like networks, a computationally feasible Monte Carlo estimation approach is provided and demonstrated empirically to provide comparable results with the analytic method. The approach of [3] applies both to static and temporally evolving networks.

A separate approach involves continuous-time *Gaussian* transmission models [21, 28]. While a continuous-time transmission model is a better approximation for realistic settings, the approach in [21, 28] is limited by being designed for trees and extended to general graphs via a BFS (shortest-path) heuristic. Other temporal methods have been proposed that observe the contamination status of a subset of sensor nodes at user-controlled intervals invoking a Four-Metric approach [33], Monte Carlo methods [1], or analytical methods for time-varying networks [15]. A separate approach is based on time-reversal backward spreading, where link weights are set equal to travel time and not spreading propensity [35]. These methods are impractical for the foodborne disease context given the lack of temporal data on times through the network available for solving the problem, as discussed in Sect. 3.6.

Fewer approaches to source detection exist within the category of non-temporal approaches based on partial observations. A line of work based on the notion of *Jordan centrality* has led to multiple variants of a technique that chooses the source node with the shortest maximum path length over all observations, that is, the Jordan center [37]. While this method has been extended to incorporate weights along the edges[1] [38], it relies on path lengths to discriminate between sources. Furthermore, the technique is designed for tree-like networks; for application to general topologies an alternate procedure based on closeness centrality (i.e., counting the sum of the shortest path to each observation) is proposed.

In addition, many of the methods based on partial observations in both the temporal and non-temporal categories are developed in the context of contagion spreading models [1–3, 20, 33, 37, 38], and are therefore inapplicable in the case of the supply network-mediated diffusion process of foodborne disease spread. As explained in Sect. 3.2, network-based diffusion is the mechanism moving contamination forward through the supply chain, which represents a different dynamics than contaminated individuals changing infection state and growing the infection. If contamination

---

[1]In the contagious disease context, normalized weights can be interpreted as heterogeneous infection probabilities.

models are applied to foodborne disease spread, the extremely low infectivity rate and recovery rates would mean that the disease would die out, and the forward diffusion of already-contaminated items would not be accounted for.

Another line of work in the category of non-temporal approaches involves a measure of *Effective Distance* on a network [6]. The Effective Distance method is developed for identifying the source of infectious disease outbreaks spreading through global mobility networks and is therefore devised in the framework of metapopulation reaction–diffusion models. However, it does not depend explicitly on the infection quantities, but only on flow transitions between nodes. It is therefore applicable to network-diffusion-only type processes such as foodborne disease and has been evaluated in application to the 2011 outbreak of EHEC in sprouts [22].

The method is based on the concept that the trajectory of a particle diffusing through a network will primarily follow the shortest, highest probability path to any other node. The true source of an outbreak should therefore be the node that exhibits the set of shortest, highest probability paths to the outbreak node set. Based on this logic, the authors introduce a metric for the Effective Distance $d_{eff}(i, j)$ between two connected nodes i and j, defined such that the likelier the connection, the shorter the Effective Distance. This is given as

$$d_{eff}(i, j) = 1 - \log p_{ij}, \tag{1}$$

where p_{ij} is the probability of transiting from i to j. The effective length of a given path γ_{so} between source node s and observation node o is then defined to be the sum total of the Effective Distances of each edge $(i, j) \in \gamma_{so}$. As discussed, the concept of [6, 22] is to focus on the shortest Effective Distance path over all possible paths $\gamma_{so} \in \Gamma_{so}$ from s to o. The Effective Distance between s and o is then defined as

$$
\begin{aligned}
D_{eff}(s, o) &= \min_{\gamma_{so} \in \Gamma_{so}} \sum_{(i,j) \in \gamma_{so}} 1 - \log p_{ij} \\
&= \min_{\gamma_{so} \in \Gamma_{so}} [|\gamma_{so}| - \log P(\gamma_{so}|s)].
\end{aligned}
\tag{2}
$$

The Effective Distance of a path therefore results from a multifactorial objective function that penalizes topologically long path lengths (the $|\gamma_{so}|$ term in the minimization) while rewarding high path probabilities (the $-\log P(\gamma_{so}|s)$ term). To identify the source of an outbreak, the single shortest Effective Distance path to each observation is identified. The source is then chosen as the node that minimizes the average and variance of the shortest Effective Distance path to each observation.

As mentioned above, the Effective Distance method was designed for application to infectious disease outbreaks spreading over global mobility networks. These networks are characterized by great heterogeneity in path lengths and probabilities, meaning that spreading processes on these networks will be dominated by a small percentage of the shortest, highest probability transport connections. As expected, the Effective Distance method performs well in settings involving outbreaks of infectious disease (e.g., SARS, H1N1) spreading through global air travel networks

[6]. Nonetheless, it is a heuristic approach that considers only a single path to each observation. While this type of approximation may be justified in certain network contexts such as the global air mobility networks the method was designed for, it is not adapted for the structure of food supply networks which are characterized by homogeneity in path lengths and the existence of multiple paths of similar probability (see Sect. 3). When the method is applied to the 2011 EHEC (foodborne disease) outbreak, source identification results are less accurate and more unstable than the infectious disease case examples [6, 22].

Two recent works have addressed the single path limitation. First, Ianelli et al. [16] have developed a generalization of the Effective Distance approach to include multiple transmission routes in estimating disease arrival times. This work leverages random walk theory to analytically demonstrate that the single path approach is an approximation of more general logarithmic network-based measures. While both methods are developed in the framework of metapopulation reaction–diffusion models, only the multiple paths approach depends explicitly on the dynamical quantities of the SIR model. This generalized effective distance approach for estimating (forward) disease propagation arrival times is therefore a departure point for an improved and analytical approach to the (inverse) source detection problem for metapopulation propagation processes like infectious diseases spreading through global air traffic networks.

More recently, the source detection problem for network-diffusion-only type processes such as foodborne disease has been solved using a similar analytical approach to account for all trajectories between source and observation. The work of Horn and Friedrich [14] formulates a probabilistic model of the contamination diffusion process as a random walk on a network and derives the maximum likelihood estimator for the source location. By modeling the transmission process as a random walk, this work develops a novel, computationally tractable solution to the inverse problem that accounts for all possible paths of travel through the network. Improvements in accuracy and stability are demonstrated in comparison with the single paths approach of [6, 22], when both methods are applied to different network topologies including stylized models of food supply network structure as well as the 2011 EHEC outbreak in Germany.

5 Summary

Many existing approaches to the source detection problem cannot be implemented in the foodborne disease context because they are designed for a different purpose— identifying the source of an epidemic contagion [1–3, 20, 33, 37, 38] whereas foodborne disease is spread according to a network-mediated diffusion process, or because they require data that is not realistically available—complete observations of the contamination status of all nodes in the network [8, 11, 30, 34] or timed network data [1–3, 15, 20, 21, 28, 33, 35]. Those that are implementable are limited by unrealistic assumptions regarding the transmission process. These methods apply tree-like approximations to deal with general graphs, assuming contamination always travels from source to observations along the shortest, highest probability paths [6, 22]. While this type of approximation is justified in certain network contexts, food supply networks are not well approximated by tree structure. Moreover, these methods are by definition approximations that do not explore the full set of trajectories between each source and observation.

To address this limitation, recent work has developed a source detection approach based on a random walk transmission model that presents a computationally tractable approach to calculate the total probability of traveling between a source and each observation along all possible paths of all possible lengths [14]. The resulting approach is not only relevant for solving the source identification problem in food supply networks but also represents a methodological improvement for source identification in diffusion processes more generally.

References

1. Agaskar, A., Lu, Y.M.: A fast Monte Carlo algorithm for source localization on graphs. In: Proc. SPIE Optical Engineering Applications, San Diego, CA (2013). Art. no. 88581N
2. Altarelli, F., Braunstein, A., Dall'Asta, L., Lage-Castellanos, A., Zecchina, R.: Bayesian inference of epidemics on networks via belief propagation. Phys. Rev. Lett. **112**(11), 118701 (2014)
3. Antulov-Fantulin, N., Lančić, A., Šmuc, T., Štefančić, H., Šikić, M.: Identification of patient zero in static and temporal networks: robustness and limitations. Phys. Rev. Lett. **114**(24), 248701 (2015)
4. Balster, A., Friedrich, H.: Dynamic freight flow modelling for risk evaluation in food supply. Transport. Res. E Log. **121**, 4–22 (2019)
5. Brockmann, D., David, V., Gallardo, A.M.: Human mobility and spatial disease dynamics. Rev. Nonlinear Dyn. Complex. **2**, 1–24 (2010)
6. Brockmann, D., Helbing, D. The hidden geometry of complex, network-driven contagion phenomena. Science **342**(6164), 1337–1342 (2013)
7. Colizza, V., Barrat, A., Barthelemy, M., Valleron, A.J., Vespignani, A.: Modeling the world-wide spread of pandemic influenza: baseline case and containment interventions. PLoS Med. **4**, 1–16 (2007)
8. Comin, C.H., da Fontoura Costa, L.: Identifying the starting point of a spreading process in complex networks. Phys. Rev. E **84**(5), 056105 (2011)
9. Crowe, S.J., Mahon, B.E., Vieira, A.R., Gould, L.H.: Vital signs: multistate foodborne outbreaks – United States, 2010–2014. MMWR Morb. Mortal. Wkly. Rep. **64**, 1221–1225 (2015)
10. Finkelstein, S.N., Larson, R.C., Nigmatulina, K., Teytelman, A.: Engineering effective responses to influenza outbreaks. Ser. Sci. **7**, 119–131 (2015)
11. Fioriti, V., Chinnici, M.: Predicting the sources of an outbreak with a spectral technique. arXiv preprint arXiv:1211.2333 (2012)
12. Friedrich, H.: Simulation of Logistics in Food Retailing for Freight Transportation Analysis. Doctoral dissertation. Karlsruhe Institute for Technology, Karlsruhe (2010)
13. Hollingsworth, T.D., Ferguson, N.M., Anderson, R.M.: Will travel restrictions control the international spread of pandemic influenza? Nat. Med. **12**, 497–499 (2006)
14. Horn, A.L., Friedrich, H.: Locating the source of large-scale outbreaks of foodborne disease. J. R. Soc. Interface **16**, 20180624 (2019). https://doi.org/10.1098/rsif.2018.0624

15. Hu, Z.L., Shen, Z., Cao, S., Podobnik, B., Yang, H., Wang, W.X., Lai, Y.C.: Locating multiple diffusion sources in time varying networks from sparse observations. Sci. Rep. **8**(1), 2685 (2018)
16. Iannelli, F., Koher, A., Brockmann, D., Hšvel, P., Sokolov, I.M.: Effective distances for epidemics spreading on complex networks. Phy. Rev. E **95**(1), 012313 (2017)
17. Keeling, M.J., Eames, K.T.D.: Networks and epidemic models. J. R. Soc. Interface **2**, 295–307 (2005)
18. LeBlanc, D.I., Villeneuve, S., Beni, L.H., Otten, A., Fazil, A., McKellar, R., Delaquis, P.: A national produce supply chain database for food safety risk analysis. J. Food Eng. **147**, 24–38 (2015)
19. Lind, P.G., da Silva, L.R., Andrade, J.S., Herrmann, H.J.: Spreading gossip in social networks. Phys. Rev. E **76**, 036117 (2007)
20. Lokhov, A.Y., Meézard, M., Ohta, H., Zdeborováá, L.: Inferring the origin of an epidemic with a dynamic message-passing algorithm. Phys. Rev. E **90**(1), 012801 (2014)
21. Louni, A., Subbalakshmi, K.P.: A two-stage algorithm to estimate the source of information diffusion in social media networks. In: Proceeding IEEE Conference on Computer Communications Workshops (INFOCOM WKSHPS), Toronto, ON, pp. 329–333. IEEE, Piscataway (2014)
22. Manitz, J., Kneib, T., Schlather, M., Helbing, D., Brockmann, D.: Origin detection during foodborne disease outbreaks-a case study of the 2011 EHEC/HUS outbreak in Germany. PLoS Curr. **6** (2014)
23. Moore, C., Newman, M.E.J.: Epidemics and percolation in small-world networks. Phys. Rev. E **61**, 5678–5682 (2000)
24. Newman, M.E.J.: Spread of epidemic disease on networks. Phys. Rev. E **66**, 016128 (2002)
25. Nigmatulina, K.R., Larson, R.C.: Living with influenza: impacts of government imposed and voluntarily selected interventions. Eur. J. Oper. Res. **195**(2), 613–627 (2009)
26. Pastor-Satorras, R., Vespignani, A.: Epidemic spreading in scale-free networks. Phys. Rev. Lett. **86**, 3200–3203 (2001)
27. Pastor-Satorras, R., Castellano, C., Van Mieghem, P., Vespignani, A.: Epidemic processes in complex networks. Rev. Mod. Phys. **87**(3), 925 (2015)
28. Pinto, P.C., Thiran, P., Vetterli, M.: Locating the source of diffusion in large-scale networks. Phys. Rev. Lett. **109**(6), 068702 (2012)
29. Pouillot, R., Lubran, M.B., Cates, S.C., Dennis, S.: Estimating parametric distributions of storage time and temperature of ready-to-eat foods for US households. J. Food Prot. **73**(2), 312–321 (2010)
30. Prakash, B.A., Vreeken, J., Faloutsos, C.: Efficiently spotting the starting points of an epidemic in a large graph. Knowl. Inf. Syst. **38**(1), 35–59 (2014)
31. Riley, S.: Large-scale spatial-transmission models of infectious disease. Science **316**, 1298–1301 (2007)
32. Scallan, E., Hoekstra, R.M., Angulo, F.J., Tauxe, R.V., Widdowson, M.A., Roy, S.L., Jones, J.L., Griffin, P.M.: Foodborne illness acquired in the United States – major pathogens. Emerg. Infect. Dis. **17**(1), 7–15 (2011)
33. Seo, E., Mohapatra, P., Abdelzaher, T.: Identifying rumors and their sources in social networks. In: Proceeding SPIE Defense, Security, and Sensing, Baltimore, MD, USA (2012)
34. Shah, D., Zaman, T.: Rumors in a network: who's the culprit? IEEE Trans. Inf. Theory **57**(8), 5163–5181 (2011)
35. Shen, Z., Cao, S., Wang, W.X., Di, Z., Stanley, H.E.: Locating the source of diffusion in complex networks by time-reversal backward spreading. Phys. Rev. E **93**(3), 032301 (2016)
36. Zhang, Z.K., Liu, C., Zhan, X.X., Lu, X., Zhang, C.X., Zhang, Y.C.: Dynamics of information diffusion and its applications on complex networks. Phys. Rep. **651**, 1–34 (2016)
37. Zhu, K., Ying, L.: Information source detection in the SIR model: a sample path based approach. In: Proceedings of Information Theory Applications Workshop (ITA), San Diego, CA, pp. 1–9. IEEE, Piscataway (2013)
38. Zhu, K., Ying, L.: A robust information source estimator with sparse observations. Comput. Soc. Net. **1**(1), 1 (2014)

Part III
Theoretical Models and Applications

Network Representation Learning Using Local Sharing and Distributed Matrix Factorization (LSDMF)

Pradumn Kumar Pandey

Abstract Vector embedding over a real network is considered as feature learning of nodes of the network which is utilized in many downstream machine learning applications such as link prediction. A network of size n can be represented as a collection of n vectors (feature vectors) of dimension d ($\ll n$) which have encoded structural and spectral information of the associated network. These feature vectors can be used in two ways: first, in the extraction of existing links and other higher order structural or functional relations among the nodes of the network and second, in the prediction of the structural evolution of the network in near future. It is observed that matrix factorization based vector embedding algorithms are able to learn more informative feature vectors but scalability is a major bottleneck due to memory and computationally intensive task.

In this paper, we present a novel distributed algorithm to learn feature vectors. It is considered that a node stores one feature vector of one of its neighbours known as shared vector, along with its own feature vector. And the learning of feature vector of a node includes only feature vectors and shared vectors stored in its neighbourhood only. Feature vectors get updated during the learning, so is shared vectors. Hence, a local sharing phenomenon leads to sharing of global information dynamically. The proposed distributed algorithm learns matrix factorization of a given network in which a node only utilizes the information available at its neighbouring nodes and connected nodes exchange feature vectors dynamically. Thus, the proposed algorithm doesn't have the limitation on its scalability. The performance of the proposed distributed algorithm for network representation learning (NRL) is evaluated for the learning of first-order proximity, spectral distance, and link prediction. The proposed network representation learning algorithm outperforms the existing state-of-the-art NRL algorithms such as node2vec, deep walk, and edge-based matrix factorization.

Keywords Distributed matrix factorization · Local sharing · Graph embedding · First-order proximity · Link prediction · Network representation learning

P. K. Pandey (✉)
Indian Institute of Technology Roorkee, Roorkee, India

© Springer Nature Switzerland AG 2019
F. Ghanbarnejad et al. (eds.), *Dynamics On and Of Complex Networks III*,
Springer Proceedings in Complexity, https://doi.org/10.1007/978-3-030-14683-2_8

1   Introduction

Networks are ubiquitous in real world which are the graphical representations of the complex systems such as biological systems, transportation, Internet, and social systems [15, 19, 26]. Network representation of a complex system is known to facilitate useful information that can be utilized for the better control and understanding of the system [11, 15]. Classification of nodes, link prediction, and recommendation are commonly studied problems in network science and machine learning which have a wide range of applications [5]. Recently low-dimensional vector embedding methods are used extensively to learn feature vectors corresponding to nodes of a network and the learned feature vectors are utilized in classification and link prediction tasks [14].

The aim of the network embedding in low-dimensional subspace is to provide a *meaningful representation* of each node in the form of vectors keeping the original network structure obtainable with high accuracy [14]. Due to the utility of low-dimensional graph embedding which is also known as network representation learning (NRL), in practical applications such as classification of nodes and link prediction, recently many attempts have been made which can be broadly characterized in two classes [14]: one is based on matrix factorization and the other is based on random walks which utilize the linear sequence learning methods from Natural-Language-Processing (NLP) [18]. In matrix factorization methods, there are two branches in which first considers the methods based on the factorization of the matrix of the low-level structure of the underlying network such as adjacency matrix and Graph-Laplacian [3], second branch deals with the methods which utilize the factorization of matrices obtained by higher order of adjacency matrix [21].

Multiple issues of existing NRL algorithms such as scalability, preservation of higher order structure and proximity have been already reported in the literature, still, these methods provide a fundamental foundation to NRL. It is accepted that matrix factorization based vector embedding algorithms are able to learn more informative features but scalability is a major bottleneck due to memory and computationally intensive task [29].

In this paper, we adopt the framework of networked multi-agent systems in which local dynamics leads to the collective behaviour of the whole networked system. We develop an algorithm which can be deployed in the form of a distributed system and have low space-and-computation complexity. The key to the proposed algorithm is the flow of feature vectors during the leaning processes which makes the learning of zeros of the adjacency matrix easier under the local access. The performance of the proposed algorithm is tested for the link extraction, spectral distance, and link prediction over some real networks and computer generated networks which are obtained under different models. The obtained feature vectors under the proposed NRL algorithm outperform the state-of-the-art NRL algorithms such as node2vec (**n2v**) [12], Deepwalk (**DW**) [25], and Gaussian matrix factorization algorithm (**MF**) [1].

The organization of the rest of the paper is as follows: The next section discusses the timeline development of the network representation learning methods. Section 3 is dedicated to explaining our approach to achieve meaningful and more effective feature learning or graph embedding over the considered networks. In Sect. 4, we provide a numerical analysis of the proposed learning algorithm and performance comparison with the other considered leading algorithms of NRL. In Sect. 5, we test the applicability of the proposed NRL learning algorithm in link prediction and finally conclude in Sect. 6 including potential future directions.

2   Related Work

Network representation learning can be viewed as feature learning and dimension reduction because the learning process is based on encoding and decoding which preserves the structure of the networks and embedded in low-dimensional space. Several theories of matrix algebra, statistics, random walk, and NLP have been utilized to develop NRL algorithms with specific utilities such as classification, higher order proximity preservation, and link prediction.

Matrix factorization based NRL algorithms utilize adjacency matrix or its derivatives such as Laplacian, and similarity matrices to learn vector embeddings corresponding to the considered network [1, 21]. In [21], similarity matrix $M = \beta(I - \beta A)^{-1}A$ is considered for matrix factorization using singular vector decomposition (SVD) with the claim that it is able to preserve higher order proximity in NRL. In [8], series of A^k are used for the same as we know that in A^k ijth entry represents the number of walks of length k. But as we include the higher order of A, the complexity of the algorithm increases and reduces the scalability and applicability of the learning process.

In other direction, SkipGram model for language learning is used for NRL with the combination of random walk. A random walk is a connected sequence of vertices which is assumed as a sentence and SkipGram is applied over it. Unbiased random walk (deep walk) and biased random walk are considered in [25] and [12], respectively, for NRL with the claim that the performance of the biased random walk is better as compared to deep walk which is an unbiased random walk. In random walk based learning algorithms, learner tries to learn the association between nodes of higher order proximity. There is an indirect big intersection between the two considered classes of NRL algorithms, most of them utilize a higher order of proximity among nodes to learn feature vectors. For more information and literature review, see [29].

3   Local Sharing and Distributed Matrix Factorization

Consider a network $G(V, E)$ of $|V| = n$ nodes and $|E| = m$ edges, where V is the node set and $E \subseteq V \times V$ is the edge set of the network. A is the adjacency matrix associated with the network G. Let $S \in \mathbb{R}^{d \times n}$ be the desired subspace to be learned such that $v_i \in S$ is the learned vector corresponding to the node i.

Here, we first do an analysis of the existing [1] regularized Gaussian matrix-factorization (MF) which is based on learning over edge set only, considering objective function $f(S, E, \alpha)$ given as

$$ f(S, E, \alpha) = \frac{1}{2} \sum_{(i,j) \in E} (v_i^T v_j - 1)^2 + \frac{\alpha}{2} \sum_i ||v_i||^2. \tag{1} $$

$$ \frac{\partial f}{\partial v_i} = \sum_{j \in N_i} (v_i^T v_j - 1) v_j + \alpha v_i, \tag{2} $$

where N_i is the set of neighbouring nodes of node i, and α is a constant. The complexity of the matrix factorization proposed in [1] is of $O(n)$ but exclusion of zeros in the learning leads to information loss. Here, we propose a more informative and efficient matrix factorization algorithm having order of complexity similar to the algorithm **MF**.

We include learning of zeros as well to get more informative feature vectors and the explanation is motivated by the concept of vector space and its null-space. The proposed algorithm to learn feature vectors through matrix factorization is distributed and efficient in its running cast which makes it scalable and improves its applicability.

Here, we provide the details of our proposed feature learning process (NRL) for a node and each node follows the same. Let S_i be the set of vectors which is formed by the vectors v_j corresponding to direct neighbours of node i and called as the vector space of node i and N_i is another set of vectors obtained by $S \backslash \{S_i \cup v_i\}$ which is called as null-space of node i. We consider $v_i^T v_j$ as decoder for the learning process which implies that $v_i^T v_j$ should be 1 if nodes i and j are connected otherwise 0. It provides that for each node i, v_i should be within or close to S_i and perpendicular or close to orthogonal to N_i. The whole process for each node is all about learning two subspaces which are almost perpendicular. The learning of perpendicular subspaces provides more accurate collective learning. Now one may ask why not perfect perpendicular? because the network is connected so it is impossible to get perfect perpendicular subspaces for low-dimensional graph embedding.

The objective function to learn feature vector corresponding to node i according to the above-explained learning process is defined as

$$ \mathcal{E}_i = \frac{1}{2|S_i|} \sum_{j \in N_i} (v_i^T v_j - 1)^2 + \frac{1}{2|N_i|} \sum_{j \notin N_i} (v_i^T v_j)^2, \tag{3} $$

where set N_i contains direct neighbours of node i. $|S_i| = d_i$ is the size of vector space which is equal to the degree of node i and $|\mathcal{N}_i| = n - d_i - 1$ is the size of null-space which increases the space-and-computational complexity of the learning process and makes the matrix factorization infeasible for large networks. To tackle this issue, instead of considering the whole null-space $\mathcal{N}_i$, we consider a subset of null-vectors $(\subset \mathcal{N}_i)$ and the size of the selected subset of null-vectors is the size of the corresponding vector space S_i. The detail of the selection of null-vectors is given later in this section. This way the runtime complexity of the proposed algorithm would be of the order of the algorithm presented in [1] with more informative feature learning.

Note Here the meaning of the size of null-space or vector subspace is in the context of the number of column vectors in $\mathcal{N}_i(\mathcal{N})$ or $S_i(S)$, respectively.

Gradient descent method is adopted for the learning of vector subspace S. In each iteration of the learning, a vector v_i changes with the rate δ in the direction of the vector $\nabla \mathcal{E}_i$ which is given by

$$\nabla \mathcal{E}_i = \frac{1}{|S_i|} \sum_{j \in N_i} (v_i^T v_j - 1)v_j + \frac{1}{|\mathcal{N}_i|} \sum_{j \notin N_i} (v_i^T v_j)v_j, \tag{4}$$

Now the question is that how to get access to the null-space of a given node, effectively?

3.1 Execution of Local Sharing Distributed Matrix Factorization (LSDMF)

In this algorithm, a node i stores two vectors v_i and $x_i \in S_i$. Similarly node j stores two vectors v_j and $x_j \in S_j$. When node i gets access to its neighbourhood for vectors v_js and x_js, it learns its complete subspace S_i and partial null-space $(\subset \mathcal{N}_i)$, $(x_j = v_k$ in Fig. 1). Here, the size of the partial null-space is fixed but its orientation is not. Why? After gets updated in each step, node i shares its updated vector v_i to one of its neighbours, selected randomly each time, let's say node j in Fig. 1. Now $x_j = v_i$ which can be used to learn null-space of node k in coming iteration of the learning. Each node follows these steps in each iteration. The philosophy of the proposed learning process is as follows: A node learns its feature vector v_i and shares it with one of its neighbours $(x_j, $ if $(i, j) \in E)$. Hence, $x_j = v_i$, which would be a vector of the null-space of the second neighbours of node i, let's say k, and they would be able to learn partial null-space during the corresponding iterations of the proposed algorithm. The process of sharing the updated vector is random. So, partial null-space gets changed in each iteration but remains within the null-space of the node i. This way all the nodes in the network share null-space locally. Here, access of null-space under dynamic local sharing is similar to sample the

Fig. 1 Pictorial
representation of local
sharing of feature vectors

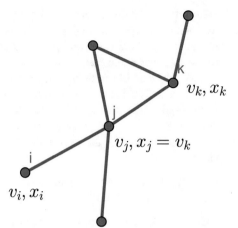

vector x_j at random from the neighbours of node j that are not also neighbours of node i during the learning at i. And dynamic sharing also helps in avoiding overfitting due to the existence of the local clustering. Hence, each node needs to access its direct neighbourhood only and the proposed algorithm works in a distributed manner. Now, the direction vector to update the feature vector v_i for the proposed local sharing distributed matrix factorization algorithm would be

$$\nabla \mathcal{E}'_i = \frac{1}{d_i} \left(\sum_{j \in N_i} (v_i^T v_j - 1) v_j + \sum_{j \in N_i} (v_i^T x_j) x_j \right), \tag{5}$$

and the pseudo code of the algorithm is given in Algorithm 1.

Algorithm 1 LSDMF

Input: A, d.
Output: $\mathcal{S}$.
 1: **procedure** SUBSPACE LEARNING
 2: Initialize $X \in \mathbb{R}^{d \times n}, \mathcal{S} \in \mathbb{R}^{d \times n}$.
 3: $x_i \leftarrow \frac{X(:,i)}{\|X(:,i)\|}, v_i \leftarrow \frac{\mathcal{S}(:,i)}{\|\mathcal{S}(:,i)\|}$.
 4:
 5:      **for** $t = 1 : T$ **do**   ▷ Learning time (iteration number).
 6:          **for** $i = 1 : n$ **do**
 7:              $v_i \leftarrow v_i - \delta \times \nabla \mathcal{E}'_i$,   ▷ δ is learning rate.
 8:              $j \in N_i$ ▷ Randomly selected from direct neighbours.
 9:              $x_j \leftarrow v_i$,   ▷ Local sharing.
10: Local sharing: In each iteration of the node i, node j is selected randomly from the direct
    neighbours of the node.
        $\mathcal{S}(:, i) \leftarrow v_i$.

---

4 Experiments and Results

In this section, we provide a numerical analysis of the proposed LSDMF over diverse data sets.

Datasets We consider nine different networks including biological network, power-grid network, social networks, and collaboration networks to evaluate the novelty and performance of the proposed subspace learning algorithm, LSDMF. We include different data sets of real networks of various sizes including Zachary's karate club network (Karate) [20], the network of interactions between major characters in the novel Les Miserables (Lesmis) [20] by Victor Hugo, Football network [28], Email network [13], Protein–Protein Interaction network (PPI) [6], largest component of a network of collaborations between physicists who conduct research on networks (CbN) [17], Power-Grid network (PGN) [27], citation network (Ca-HepTh) [17], PGP network [4], Facebook and AstroPH [12]. All the considered data sets are publicly available [16]. The details of the data sets are given in Table 1.

We also include synthetic networks obtained under the models CDPAM with parameter value ($\beta = 0.6$) [22], BA [2], WS with rewiring probability $p = 0.01$ [27], NRM with parameters ($\beta = 0.1$ and $p = 0.3$) [23], and FFM with parameter value of forward burning probability $= 0.3$, and the backward burning ratio $r = 0.35$ [17] to evaluate the applicability and superiority of the proposed LSDMF for the learning of graph embeddings under different graph generation scenarios.

Table 1 Characteristics of the networks used for validation

Network/dataset	n	m	Q
Karate [20]	34	78	0.4188
Lesmis [20]	77	254	0.5556
Football [28]	115	613	0.6046
Email [13]	1133	5451	0.5406
PPI [6]	2361	6646	0.5894
CbN [17]	4158	13,425	0.8191
PGN [27]	4941	6594	0.7701
Ca-HepTh [17]	8638	24817	0.7240
PGP [4]	10,680	24,316	0.8459
Facebook [17]	4039	88,234	0.7774
AstroPH [17]	18,533	396,160	0.8123
CDPAM [22]	5000	20,000	0.1821
BA model [2]	5000	20,000	0.2133
WS Model [27]	5000	25,000	0.2764
NRM [23]	5000	19,133	0.7342
FFM [17]	5000	15,583	0.7783

n number of nodes, m number of edges, and Q modularity of the network

These models include random networks which exhibit scale-free degree distribution (CDPAM, BA, NRM, and FFM) and binomial degree distribution (WS). CDPAM, BA, and WS are the models without community structure, and NRM and FFM have non-vanishing local clustering.

Metrics We use two metrics for measuring the performance of different NRL algorithms by quantifying the distance between two networks, the original network and the reconstructed network. First is the well-known *spectral distance* [7]. Let A and C be the adjacency matrices of original and reconstructed networks, both having n nodes. Also, let $\lambda = [\lambda_1, \geq \lambda_2, \geq, \ldots, \geq \lambda_n]$ and $\mu = [\mu_1, \geq \mu_2, \geq, \ldots, \geq \mu_n]$ be the sorted list of eigenvalues of A and C, respectively. The spectral distance between A and C is defined as:

$$d(A, C) = \sum_{i=1}^{n} |\lambda_i - \mu_i|$$

Second distance metric used is the *fraction of wrongly reconstructed edges* between original adjacency matrix A and its reconstruction C [24], which is given by

$$\Delta E = \frac{1}{2} \frac{\mathbf{1}^T abs(A - C)\mathbf{1}}{\mathbf{1}^T A \mathbf{1}},$$

where $[abs(A)]_{ij} = |A_{ij}|$ and $\mathbf{1}$ is all-one vector of length n.

4.1 Simulation

We consider 3 NRL algorithms in which two are based on random walk node2vec (**n2v**) [12] and Deepwalk (**DW**) [25], and an edge-based matrix factorization, to compare the performance of our proposed LSDMF. Matrix factorization algorithm, noted as **MF**, proposed in [1] is considered as a baseline matrix factorization algorithm which considers only edge set for learning purpose. Performance deference between **MF** and LSDMF (see columns 4 and 5 in Table 2) directly makes us notice the importance of the consideration of null-space. LSDMF exhibits better performance as compared to **n2v**, **DW**, and **MF**, see Table 2. In simulations, we consider the integer value of $\sqrt{n}$ as the dimension d, where n is the size of the considered network. In case of **n2v**, **DW**, we set the length of random walk 80, window size 10, number of walks per node 10. These are the values of common parameters for **DW** and **n2v**. The considered hyper-parameters are $p = q = 1$ in case of **DW** and $p = 1$, $q = 0.5$ in case of **n2v**. For more details of the selection of parameters of **DW** and **n2v**, see [25] and [12], respectively.

Effect of Dimension d We also investigate the effect of the dimension, d, of the learned feature vectors. Error in link extraction is considered to evaluate it. For

Table 2 Comparison of reconstruction errors for various algorithms

Method	n2v	DW	MF	LSDMF
Precision $(1 - \Delta E)$				
Real networks				
Karate	0.19	0.21	0.37	**0.64**
Lesmis	0.26	0.22	0.45	**0.65**
Football	0.54	0.58	0.33	**0.95**
Email	0.31	0.27	0.12	**0.84**
PPI	0.12	0.09	0.14	**0.65**
CbN	0.34	0.29	0.19	**0.80**
PGN	0.33	0.37	0.13	**0.88**
Ca-HepTh	0.35	0.32	0.20	**0.84**
PGP	0.27	0.24	0.11	**0.53**
Synthetic networks				
CDPAM	0.42	0.43	0.16	**0.75**
BA model	0.46	0.42	0.25	**0.86**
WS model	0.52	0.55	0.18	**0.72**
NRM	0.33	0.37	0.19	**0.94**
FFM	0.54	0.49	0.13	**0.89**
Spectral distance: $d(A, C)$				
Real networks				
Karate	23.3	21.7	19.2	**13.6**
Lesmis	39.2	37.3	56.7	**28.4**
Football	151.1	78.5	75.3	**11.2**
Email	634.6	661.2	578.7	**355.3**
PPI	2504.8	2689.1	1424.0	**757.4**
CbN	2949.0	2734.4	3735.0	**823.2**
PGN	2948.4	2853.1	4928.2	**652.3**
Ca-HepTh	7173.4	4972.1	7253.7	**1808.3**
PGP	7826.4	7536.0	8972.7	**1782.3**
Synthetic networks				
CDPAM	3852.2	3922.8	4271.2	**1721.6**
BA model	1281.7	982.2	1342.2	**365.0**
WS model	723.5	1329.1	942.2	**120.0**
NRM	4283.2	3826.2	2312.5	**258.1**
FFM	3293.8	2836.3	2783.2	**657.9**

The values in bold are corresponding to the best performing algorithm among the considered NRL algorithms

different values of d, we calculate the wrongly identified links (ΔE) for the network Email. Accuracy of the network reconstruction $(1 - \Delta E)$ is plotted in Fig. 2 from which it is observed that our proposed algorithm for NRL exhibits superiority as compared to **n2v**, **DW**, and **MF**. The same observation is valid for other considered networks also.

Fig. 2 Performance of
different NRL algorithms at
different dimensions of
learned graph embeddings for
email network

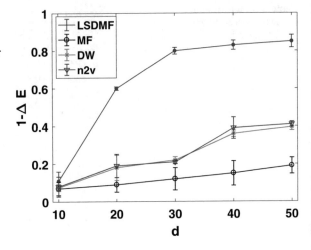

Fig. 3 Convergence of
LSDMF in Email network.
Horizontal axis represents
iteration number and vertical
axis represents accuracy in
network reconstruction

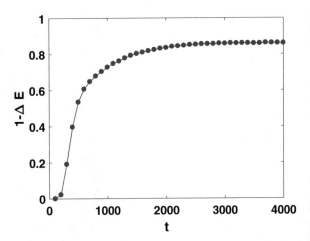

Convergence Learning of feature vectors under considered NRL algorithms, **n2v**, **DW** uses standard machine learning function *Softmax* and **MF** uses stochastic gradient descent method over links for learning. We also utilize the stochastic gradient descent considering links and randomly selected non-links. LSDMF selects non-links randomly in each iteration of learning and utilizes defined local sharing scheme. Convergence of the proposed LSDMF is shown in Fig. 3. Accuracy of the reconstruction of the underlying network $(1 - \Delta E)$ is plotted for different values of learning time t in Fig. 3. We get similar results for other data sets also.

5 Application: Link Prediction

In this section, we perform link prediction task utilizing the obtained feature vectors under the different considered NRL algorithms. The task of link prediction is evaluated over the variety of data sets considered in this paper including Facebook and AstroPH which are used for the same in [12]. We follow the standard experimental setup, outlined in [10], evaluate the performance of the learning quality of the proposed NRL algorithm and other state-of-the-art methods. In our experiments, 30% links are selected randomly and deleted from the input real network. After deletion of links, the remaining network is used for learning the graph embedding or feature vectors. Next, standard literature [10] is followed for link prediction problem. From the earlier deleted set of links, 20% random links are selected as positive examples and the equal number of negative examples (non-links) which are also selected randomly to train the logistic regression model for link prediction. Remaining 10% of deleted links and the equal number of random non-links are used for testing the performance of the trained model. We calculate F1-score which is reported in Table 3. All the reported numbers are averaged over 10 random runs of the above experiment. Here, learned graph embedding is not directly used as input to the logistic regression model. Two vectors can be close to each other either due to angular distance or due to Euclidean distance. Let v_i and v_j be the vectors learned corresponding to nodes i and j, respectively. We consider dot product ($v_i^T v_j$) which resembles angular distance and 2-norm of difference vector ($||v_i - v_j||_2$) which is the Euclidean distance between the concerned vectors. These two distances between any two learned vectors in the network are used as input features for training and testing the logistic regression based predictor.

Numerical results of link prediction are reported in Table 3 from which we observe that feature vectors learned under our proposed algorithm are more informative and have better performance in the link prediction as compared to the other considered graph embeddings obtained under **n2v**, **DW** and **MF**. The performance of the proposed NLR algorithm over diverse data sets justifies the wide applicability of LSDMF.

Table 3 F1 values for link prediction using logistic regression utilizing node embedding by different NRL algorithms

Networks/methods	n2v	DW	MF	LSDMF
Football	0.69	0.65	0.58	**0.78**
Email	0.53	0.46	0.61	**0.76**
PPI	0.45	0.42	0.59	**0.73**
CbN	0.57	0.61	0.65	**0.85**
ca-HepTh	0.57	0.49	0.61	**0.81**
PGP	0.58	0.55	0.51	**0.72**
Facebook	0.61	0.65	0.71	**0.91**
AstroPH	0.41	0.45	0.76	**0.85**

The values in bold are corresponding to the best performing algorithm among the considered NRL algorithms

6   Conclusion and Future Work

In this paper, a novel algorithm for efficient matrix factorization is proposed which utilizes a local sharing of graph embedding or feature vectors during the learning process and does not require the complete visibility of the network. Only local sharing of feature vectors and locally implemented objective function leads to collective learning over the network. The proposed algorithm is able to learn more informative graph embedding which outperforms the state-of-the-art algorithms such as node2vec and Deepwalk in link prediction task, and first-order proximity preservation. The proposed matrix factorization is more informative and accurate as compared to the edge-based matrix factorization without increasing the order of computation complexity.

Apart from link prediction, there are other complex problems in which NRL can provide better and effective solutions. Few of them are listed below:

Clique Finding Higher order proximity preservation in networks is highly desirable [21] for many applications such as maximal clique finding in biological networks. We can get a number of applications of maximal clique finding problem in literature [9]. Clique, motif, dense sub-graphs, and communities are higher level structures present in real networks. Minimum perturbation of the higher level structures is a prime requirement of graph embedding in low-dimensional subspace while decoding of a network from its learned vector subspace.

Diffusion or Search Path Learning Real networks are so large that it is difficult to get access to the whole network in a single sight. In such conditions, searching and controlling the network becomes difficult. As we observe that only local learning is sufficient to learn the global structure of a node (links and non-links). Learning effective search paths in a large network is a potential problem to control diffusion processes and it can be learned using NRL.

Community Finding Community finding under NRL can be considered as a clustering problem. Learned feature vectors are simple data points and learning strategy based on Euclidean or angular distances learns feature vectors in such a way that densely connected group of nodes have close feature vectors.

References

1. Ahmed, A., Shervashidze, N., Narayanamurthy, S., Josifovski, V., Smola, A.J.: Distributed large-scale natural graph factorization. In: Proceedings of the 22nd International Conference on World Wide Web, pp. 37–48. ACM, New York (2013)
2. Barabási, A.L., Albert, R.: Emergence of scaling in random networks. Science **286**(5439), 509–512 (1999)
3. Belkin, M., Niyogi, P.: Laplacian eigenmaps and spectral techniques for embedding and clustering. In: Advances in Neural Information Processing Systems, pp. 585–591 (2002)
4. Boguná, M., Pastor-Satorras, R., Díaz-Guilera, A., Arenas, A.: Models of social networks based on social distance attachment. Phys. Rev. E **70**(5), 056122 (2004)

5. Bornholdt, S., Schuster, H.G.: Handbook of Graphs and Networks: From the Genome to the Internet. Wiley, Weinheim (2006)
6. Bu, D., Zhao, Y., Cai, L., Xue, H., Zhu, X., Lu, H., Zhang, J., Sun, S., Ling, L., Zhang, N., et al.: Topological structure analysis of the protein–protein interaction network in budding yeast. Nucleic Acids Res. **31**(9), 2443–2450 (2003)
7. Butler, S., Chung, F.: Spectral graph theory. In: Handbook of Linear Algebra, p. 47. CRC Press, Boca Raton (2006)
8. Cao, S., Lu, W., Xu, Q.: Grarep: learning graph representations with global structural information. In: Proceedings of the 24th ACM International on Conference on Information and Knowledge Management, pp. 891–900. ACM, New York (2015)
9. Cazals, F., Karande, C.: A note on the problem of reporting maximal cliques. Theor. Comput. Sci. **407**(1–3), 564–568 (2008)
10. De, A., Bhattacharya, S., Sarkar, S., Ganguly, N., Chakrabarti, S.: Discriminative link prediction using local, community, and global signals. IEEE Trans. Knowl. Data Eng. **28**(8), 2057–2070 (2016)
11. Ferber, J.: Multi-Agent Systems: An Introduction to Distributed Artificial Intelligence, vol. 1. Addison-Wesley, Reading (1999)
12. Grover, A., Leskovec, J.: node2vec: scalable feature learning for networks. In: Proceedings of the 22nd ACM SIGKDD International Conference on Knowledge Discovery and Data Mining, pp. 855–864. ACM, New York (2016)
13. Guimera, R., Danon, L., Diaz-Guilera, A., Giralt, F., Arenas, A.: Self-similar community structure in a network of human interactions. Phys. Rev. E **68**(6), 065103 (2003)
14. Hamilton, W.L., Ying, R., Leskovec, J.: Representation learning on graphs: methods and applications. (2017, preprint). arXiv:1709.05584
15. Jackson, M.O.: Social and Economic Networks. Princeton University Press, Princeton (2010)
16. Leskovec, J., Krevl, A.: SNAP datasets: Stanford large network dataset collection (2014). http://snap.stanford.edu/data
17. Leskovec, J., Kleinberg, J., Faloutsos, C.: Graph evolution: densification and shrinking diameters. ACM Trans. Knowl. Discov. Data **1**(1), 2 (2007)
18. Levy, O., Goldberg, Y.: Dependency-based word embeddings. In: Association for Computational Linguistics ACL, vol. 2, pp. 302–308 (2014)
19. Newman, M.E.: Complex systems: a survey. (2011, preprint). arXiv:1112.1440
20. Newman, M.E., Girvan, M.: Finding and evaluating community structure in networks. Phys. Rev. E **69**(2), 026113 (2004)
21. Ou, M., Cui, P., Pei, J., Zhang, Z., Zhu, W.: Asymmetric transitivity preserving graph embedding. In: Knowledge Discovery and Data Mining (KDD), pp. 1105–1114 (2016)
22. Pandey, P.K., Adhikari, B.: Context dependent preferential attachment model for complex networks. Phys. A Stat. Mech. Appl. **436**, 499–508 (2015)
23. Pandey, P.K., Adhikari, B.: A parametric model approach for structural reconstruction of scale-free networks. IEEE Trans. Knowl. Data Eng. **29**(10), 2072–2085 (2017)
24. Pandey, P.K., Badarla, V.: Reconstruction of network topology using status-time-series data. Phys. A Stat. Mech. Appl. **490**, 573–583 (2018)
25. Perozzi, B., Al-Rfou, R., Skiena, S.: Deepwalk: online learning of social representations. In: Proceedings of the 20th ACM SIGKDD International Conference on Knowledge Discovery and Data Mining, pp. 701–710. ACM, New York (2014)
26. Scott, J.: Social Network Analysis. Sage, Thousand Oaks (2017)
27. Watts, D.J., Strogatz, S.H.: Collective dynamics of 'small-world' networks. Nature **393**(6684), 440 (1998)
28. White, S., Smyth, P.: A spectral clustering approach to finding communities in graphs. In: Proceedings of the 2005 SIAM International Conference on Data Mining, pp. 274–285. SIAM, Philadelphia (2005)
29. Zhang, D., Yin, J., Zhu, X., Zhang, C.: Network representation learning: a survey. IEEE Xplore Digital Library (2018)

The Anatomy of Reddit: An Overview of Academic Research

Alexey N. Medvedev, Renaud Lambiotte, and Jean-Charles Delvenne

Abstract Online forums provide rich environments where users may post questions and comments about different topics. Understanding how people behave in online forums may shed light on the fundamental mechanisms by which collective thinking emerges in a group of individuals, but it has also important practical applications, for instance, to improve user experience, increase engagement or automatically identify bullying. Importantly, the datasets generated by the activity of the users are often openly available for researchers, in contrast to other sources of data in computational social science. In this survey, we map the main research directions that arose in recent years and focus primarily on the most popular platform, Reddit. We distinguish and categorize research depending on their focus on the posts or on the users and point to different types of methodologies to extract information from the structure and dynamics of the system. We emphasize the diversity and richness of the research in terms of questions and methods and suggest future avenues of research.

Keywords Online communities · Stochastic models · Discussion trees

A. N. Medvedev
naXys, Université de Namur, ICTEAM, Université catholique de Louvain, Louvain-la-Neuve, Belgium

R. Lambiotte
Mathematical Institute, University of Oxford, Oxford, UK

naXys, Université de Namur, Namur, Belgium
e-mail: renaud.lambiotte@maths.ox.ac.uk

J.-C. Delvenne (✉)
ICTEAM and CORE, Université catholique de Louvain, Louvain-la-Neuve, Belgium
e-mail: jean-charles.delvenne@uclouvain.be

© Springer Nature Switzerland AG 2019
F. Ghanbarnejad et al. (eds.), *Dynamics On and Of Complex Networks III*,
Springer Proceedings in Complexity, https://doi.org/10.1007/978-3-030-14683-2_9

1   Introduction

Understanding the dynamics and structure of human communication is a central research theme in computational social science. The increasing availability of digital traces of human interactions has allowed to quantify, at a large scale, a variety of phenomena. For instance, phone call logs led to the identification of the burstiness of human communication, typically organized into "periods" of short intensive communication followed by long periods of silence [32]; Facebook and email data helped to confirm the smallness of the world, i.e. the typical network distance between people is disproportionally small as compared to its size [4]; tweet messages led to studies to uncover the mechanisms leading to information cascades [65]; etc. If early works initially focused on one-to-one communication, the emergence of new communication channels, such as Twitter or online forums, has opened the possibility to study collective discussions [2].

Collective discussions have not been invented by new media. As such, they have been and remain a major way for exchanging opinions and for producing collective decisions. Online forums provide a venue where Internet-goers post questions or comments, which may, or may not, trigger discussions from other members of the community. Understanding how people behave in online forums has important theoretical implications, to improve our understanding of collective thinking, but also practical applications, to improve user experience, increase engagement or facilitate the democratic process [1]. The purpose of this chapter is to provide an overview of the academic research on online discussion platforms, or online forums, and to bring together the variety of research questions considered in the literature. Most of our attention is dedicated to the self-proclaimed "front page of the Internet" [50]—the website Reddit (REDDIT.COM)—which is the largest online discussion forum in the world as of today. Note that several other online discussion platforms have a similar architecture and have also been studied, for instance, in comparative studies; they include Digg, Hacker News, Slashdot, Epinions, Meneame, Barrapunto and even Wikipedia.

The rest of this chapter is organized as follows: Section 2 presents the datasets that can be extracted from Reddit and have been widely used by researchers. Academic studies are then divided according to their primary focus on the post or the users and are presented in Sects. 3 and 4, respectively. We conclude with a discussion and perspectives for future research.

2   The Reddit Dataset

Reddit (launched in 2005) is a social news aggregation, web content rating and discussion website, ranked as #6 most visited website in the world with 234 million unique users (as of February 2018).[1] A schematic structure of Reddit is illustrated

---

[1] https://en.wikipedia.org/wiki/Reddit.

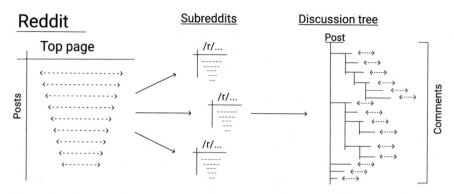

Fig. 1 The schematic structure of the Reddit platform. The entry point is the top page of Reddit, which is feed of posts from the subreddits followed by a registered user (or from all subreddits, for an anonymous user) and ranked according to votes and posts' age. The user may further proceed to the top page of a specific subreddit, where the feed narrows down to only posts from a chosen subreddit. Each post can be upvoted or downvoted and has an attached section of comments. The comments are structured as a rooted tree by the reply-to relation to other comments or the post itself

in Fig. 1. Registered users submit posts that contain a title, an external link or a self-written piece of content, which immediately become available to the whole audience of Reddit for voting and commenting. The voting system permits only registered users to upvote (give a positive $+1$ vote) or to downvote (give a negative -1 vote) on posts and comments. Comments form a *discussion tree*, which can be described as a rooted tree, where the root is a designated node representing the post itself and each other node represents a comment. There is a link between two nodes if there is a "reply-to" relation between them.

The huge posting space of Reddit is divided into subreddits—*self-created* communities of users, united by a certain topic. Every submitted post has subreddit name as an intangible attribute. Each subreddit and Reddit itself has a so-called "top page"—the feed where post titles with voting and commenting links are delivered to users. Two factors influence the post's ranking position there: (1) time and (2) voting *score*, or otherwise called *karma*, which is basically the difference between upvotes and downvotes. High score posts have a higher chance of appearing at the top page. However with time, newer information replaces the older in the feed. Users can follow subreddits, but not other users, which constitutes the main distinction with social network platforms, like Facebook or Twitter, where users follow a person and not a content. Other platforms have a similar structure. For example, Slashdot (launched in 1997) is made of news stories, together with comments moderated by selected users, but not by an open voting system.[2] Only a fixed number of topic-based subsections is available. Hacker News (launched in 2007) is an online community very similar to Reddit but with only two pre-made topic subsections

---

[2]https://en.wikipedia.org/wiki/Slashdot.

(subreddits).[3] Digg (launched in 2004) acts currently as a news aggregator, but it formerly was a socially curated platform with a post submission, commenting and voting system like Reddit.[4] Meneame is a Spanish analogue of Digg, and Barrapunto is also a Spanish version of Slashdot [19].

We chose to dedicate our attention to Reddit because of the variety of ways the system is self-organized. As it is mentioned later in the chapter, such self-organization provides place for free movement, social herding, organized attacks, trolling, etc. Subreddits as topically self-identified entities create real communities in virtual space, with their own rules, language features, intrinsic rules, jokes, etc. Such annotated text corpus becomes valuable for training machine learning algorithms [60]. General availability of information brings in different dynamics of information spread in comparison to direct followers social media (e.g. Twitter), the volume of information poses the problem of missed content, or the posts that potentially could gain comments or votes, but was missed in the avalanche of other posts. Reddit allows sharing a wide variety of content. One may use Reddit data to uncover the relationships between different Internet services.

Reddit has gained a central place in the scientific literature thanks to the openness, richness and quality of its data, which allows to perform longitudinal studies of the whole system and, critically, to ensure reproducibility of the results. Jason Baumgartner, under the Reddit name Stuck_In_The_Matrix, did a tremendous amount of work when attempted to collect a full dataset of posts and comments, going back to the creation of the site [54]. The figures of this chapter have all been prepared from this dataset. For instance, basic numbers on the growth of the site and the total sizes of discussions are found in Fig. 2. His data repository also contains the data from the platform Hacker News [53].

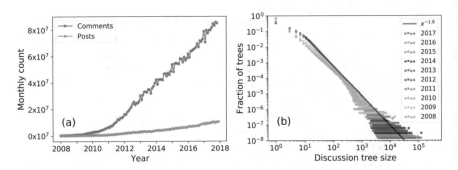

Fig. 2 The evolution of Reddit from Jan 2008 till Jan 2018: (**a**) monthly counts of posts and comments, (**b**) distribution of discussion sizes. One may notice an exponential increase in the activity counts, but the discussion size distribution follows a similar shape, close to the power-law with exponent α ranging from around 1.7 for early years to 1.9 for later

---

[3]https://en.wikipedia.org/wiki/Hacker_News.

[4]https://en.wikipedia.org/wiki/Digg.

Despite its recognized quality, one should be careful while using the dataset for research purposes. Gaffney and Matias [14] report several inconsistencies in the data. Their approach is based on the fact that a post or comment id is essentially an integer number in base36 format; thus, allegedly, all the continuous range of numbers must be present in the dataset. For example, comment and post data before 2008 appears to be hugely corrupted, having around 80% of posts missing, as well as 90% posts information from a few months data at the interface between 2009 and 2010. In total, across the time interval Jan 2006 and Feb 2016, the authors report 0.043% missing comments and 0.65% missing posts.

The risks of missampled data obviously cause commenting/posting rate distortions, missing information in user time series and possible inability to reconstruct certain discussion trees. Nevertheless the data from Jan 2008 and later is fairly consistent for detailed look and especially in large-scale studies the inconsistencies may be safely disregarded due to their smallness. The system has sustained exponential growth [50]; thus, data volume in early years is negligibly small compared to today's numbers. The proposed approach to measure missing data also raises a question of applicability, since the gaps in consecutive numbering may be due to inner technical features of the website. This may be supported by the fact that a newly published rescraped data contains the same missing values found in the data before [55]. One may use directly the Reddit API for consistency checks.[5]

Although missing data causes reasonable inconsistencies, the present data has a few important peculiarities to consider. There exist posts and comments from authors, whose accounts were deleted and the author name of the comment turns to default name "[deleted]". According to the letter appendix to [14] such comments comprise around 25% of the data. This fact imposes a greater obstacle on studies of user participation in discussions and reply networks. Another problem comes with posts or comments that were deleted by user and removed due to moderation. In the current version of the API the text body of such comments and posts is correspondingly marked as "deleted" and "removed", but it is not clear what happened in early years.

3 From the Perspective of Posts

Posts are at the heart of the platform structure and dynamics. Once posted, they may gain attention and receive feedback in the form of votes and comments, thereby obtaining a good ranking and even more attention. They may also go quickly unnoticed in the avalanche of newer posts. The next subsection is dedicated to the topic of popularity prediction in online platforms. This longstanding problem has been studied in various online systems. As first step, we thus provide a quick overview of research on the wide spectrum of online systems before concentrating on Reddit specifically.

---

[5]https://praw.readthedocs.io/en/latest.

A standard ingredient in predictive models is the incorporation of quantitative features that tends to correlate with the popularity of posts. Features can be structural, dynamical, textual and even associated with the author of the post. Initial works proposed simple statistical models based on regression, Poisson or Cox processes, but with the developments of machine learning, more elaborate methods based on neural networks have emerged.

Popularity Prediction Anyone who has ever used online social networks is familiar with the concept of "likes" and "dislikes"—the way of expressing attitude towards a piece of content on a binary scale. If "page views" have long been the dominant measure of the success of a content, more and more platforms have moved to voting systems where the number of positive votes is the measure of popularity. Discussion platforms use systems of upvotes and downvotes for different purposes, ranging from the automatic discovery of appreciated items and its delivery to a wider audience, to the moderation of discussions to protect from spam or malicious content. For these reasons, good models of popularity prediction are of interest for both content creators and platform curators.

In general, the problem of popularity prediction has been considered in various online social systems. Early studies, e.g. in YouTube and Digg, found a direct relation between content's initial popularity (in terms of views and upvotes) and its future counts [56], but more sophisticated models have been proposed since then. For instance, Lee et al. [36] have modelled the lifetime of discussions on Myspace with the Cox proportional hazard regression model. They selected the number of "risk" factors, fitted from the data for each thread, which were further used as a predictor of threads hitting a threshold number of comments. Mishne and Glance [40] analysed the corpus of comments in weblogs and the relation between weblog popularity and commenting patterns in it. Tsagkias et al. [59] analyse the corpus of comments under news stories in regional Internet news agents. The authors propose a model that predicts the commenting popularity prior to article publication in two cases: first, if there is a potential to receive comments and second, if the article receives "low" or "high" comment volume. Bandari et al. [5] also investigated if popularity of news articles can be estimated even before their posting online.

The Reddit dataset shows a proportional relation between the score of a post and the size of its discussion tree *on average*, as shown in Fig. 3. This may shed some light on general aspects of the posts' popularity; however, in each particular case, it is more important to make a more tailored estimation of submissions' score. A number of works has been dedicated to the prediction of scores on Reddit. Horne et al. [29] found a number of textual and temporal features of high score comments by considering the discussion threads from 11 popular subreddits appearing in a 6 month period of 2013. The authors proposed a machine learning model for predicting comments' score and pointed the differences in users' preferences in subreddits. In particular, they claimed that timing of the comment, its relevancy and novelty have positive impact, but stale memes or high user ranking (overall number of positive comments in a user's history) does not affect or even pushes down the average comment score. It was observed that moderation does not always impact proper behaviour in the community and may shorten the life of a discussion thread.

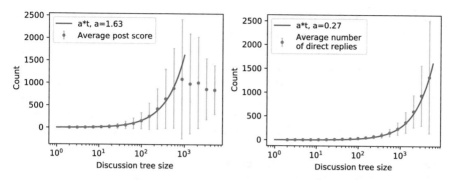

Fig. 3 Average posts' score versus discussion size (left figure) and number of direct replies to a post versus discussion size (right figure) in Reddit. The data shows that average values may be fitted with a linear trend up until a certain value. Dots represent average counts, and bars show standard deviation. The figures are based on data from year 2009, but other years show similar results

Recurrent neural networks (RNNs) have also been employed to measure community endorsement. Fang et al. [13] constructed an RNN trained to predict comment scores. Instead of controlling for the submission context, the model learns latent modes of submission context and examines how the context relates to different levels of community endorsement. On a dataset of three popular subreddits, they achieve a good performance, on average, and show that high score comments are usually harder to predict than lower ones. High-scoring comments tend to be submitted early in the discussion and the number of direct replies is not smaller than the height of its hanging discussion subtree. Low and medium score comments have a number of direct replies less than the height of a discussion subtree, indicating the presence of a further discussion. Low score comments tend to come later in the discussion overall, but also later in terms of the group of responses to a parent comment.

While structural features have been shown to be good predictors, researchers have started working on extracting textual linguistic features in order to gain predictive power. In this direction, Jaech et al. [30] have reported general improvement of machine learning classifiers in the problem of ranking comments that appear in a fixed time window in a discussion thread. Note, however, that the gain was reported to be marginal. Later, Zayats and Ostendorf [63] constructed a specific type of RNN called LSTM (long short term memory) for the same purpose of predicting comment scores. The proposed model uses structural and temporal comment features, as well as textual linguistic features of the comments. The authors achieved a slightly better performance (increase in average F1 score from 50 to 54 on average) on a dataset of three subreddits studied earlier in [13]. They found that controversial comments (that further generate a wide discussion in terms of a discussion tree) tend to be overpredicted (with a lower score than predicted) and jokes and funny comments, on the contrary, were mostly underpredicted (with a higher score than predicted). Linguistic context was found to be helpful in prediction tasks and words of underpredicted comments were aligned with comments of positive score, but words associated with overpredicted comments did not show any significant correlation. In another work, Hessel et al. [26] considered pairs of posts, submitted within a very short time interval into the same communities (to exclude timing bias) and predicted the more popular posts in those pairs. Six primarily image-sharing subreddits were selected for the study, with a set of features including textual and temporal information, but also image features assessed by the deep neural networks. The authors concluded that user-centric characteristics, e.g. previous popular submissions, and content-specific features, e.g. more complicated images and simpler titles, make a good predictor of popularity of the submission. The authors also reported an accuracy comparable to that of human classification.

The content popularity is influenced by various factors and public endorsement may not properly reflect its inherent quality [49]. This phenomenon has been studied by Stoddard [52] by means of a Poisson regression model that infers the intrinsic quality of posts from voting activity of the users. The author collected a unique dataset of users' voting time series by tracking a number of top posts on the front pages of five subreddits and the front page of Hacker News, and used data to predict final score of posts. A variable of quality was then introduced in the model parameters and was shown to correlate with the total post scores, although there were several situations when similar quality posts had different scores and vice versa. Amongst others, the mechanism of making popular more visible than less popular ones leads to a multiplicative process that increases the variance of popularity, with the effect of making a substantial fraction of the posts ignored. According to Gilbert [15], Reddit overlooks 52% of the most popular links the first time they were submitted. Lakkaraju et al. [34] explored this idea and showed that resubmissions of the same piece of content may gain more popularity than an original submission. The authors collected a dataset of image submissions between 2008 and 2013, where each image was resubmitted roughly 7.9 times.[6] Same pictures can in principle be resubmitted to different subreddits; thus, the authors employed a success metric compared to the average post score in the community. The authors also proposed a statistical model that predicts the expected score of a resubmitted picture. The model parameters include the inherent content popularity, penalties from previous success and previous submissions to other communities or to the same community twice. Overall, the study supports the hypothesis that a high quality content "speaks for itself" and determines its score. The choice of subreddit plays an important role—the model shows that the content, resubmitted to the same subreddit, in general was unlikely to be popular, as well as whether the content was previously highly rated in a popular subreddit (with a high number of visitors or subscribers). This effect gradually disappears with time, indicating forgetfulness of the audience. Titles of resubmissions were also found indicative: if the title is novel, written using subreddit-specific words and sentiment orientation, the submission

---

[6]The authors used karmadecay.com—the reverse image search tool specifically designed for Reddit.

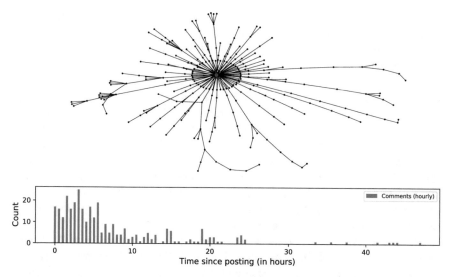

Fig. 4 Sample discussion tree of a post on Reddit with a histogram of comment arrival. Central large red node depicts the post, and comments depicted in black. The histogram presents hourly aggregation of comments arrival

has higher chances to receive positive feedback. Similarly, Glenski et al. [16, 17] found that users mostly vote on posts only after glancing at the title, without proper reading of the content or the discussion. The authors of [34] also performed an *in situ* experiment: they manually chose and resubmitted 85 images from the dataset, select a "good" and a "bad" title according to the model for each picture and post them in two different subreddits. The post scores, gathered after 1 day, show that submissions with a "good" title generated scores three times higher than the "bad" ones.

Generative Models for Discussion Trees As mentioned above, comments under the post form a rooted tree. Such trees have dynamic nature and their temporal growth reveals the dynamic of attention to the post (see an example of a tree and a histogram of comment arrival in Fig. 4). Generative models for discussion trees mostly question the tree structure of a discussion while disregarding the comments' or posts' textual features and exact timings. The reader may refer to the extensive review on generative models given by Aragón et al. [2], and here we give only a brief overview of the representative contributions.

Gomez et al. [19] considered discussion trees in four large Internet boards (Slashdot, Barrapunto, Meneame and Wikipedia) and proposed a generating model based on preferential attachment mechanism (PA model) with respect to the comment degree and the root bias. Later Gomez et al. [20] enriched this model by incorporating a notion of novelty of comments, which is represented by an exponentially decaying function of attractiveness. The model showed better results in likelihood of representing the tree structure and reproduced well the width/depth relation for discussion trees. Lumbreras et al. [37] proposed to enrich the PA model with the notion of *roles*, which are latent functions of community members. The PA model defines a set of parameters that regulate the place of attachment of a new comment. The authors suggest that this set of parameters is different for different users and propose to group them into the role sets, which are inferred accordingly. Despite this extra natural assumption, the gain in model likelihood is marginal.

Aragón et al. [3] considered the social system Meneame, where the change in discussion representation happened in 2015 from a plain list to a structured threaded view. This change was observed to have an impact on the structure of discussion trees and the authors further enrich the PA model, with a reciprocity term that captures the tendency of posting authors to reply back in the discussion. It was observed that change of the platform interface had a positive effect on reciprocity, as well as on other parameters in general.

The above mentioned models focus exclusively on the structure of discussion trees while leaving out the continuous time dynamics of the comment attraction process. Kaltenbrunner et al. [31] found that comment arrival time in Slashdot discussions fits well by double lognormal distribution, although the fitting quality depends on a circadian rhythm of the site. Based on this finding, the authors propose the prediction model that predicts the total number of comments in a discussion thread. The dynamical aspect of tree generation was first studied by Wang et al. [61], where the authors introduce a merely theoretical model for the structural and temporal evolution of discussions. The temporal evolution was described as a Lévy process with power-law interevent time distribution, when newly arriving comments were assumed to attach to the existing tree under the simple PA rule. The model was inspired by empirical observations of such discussion boards, like Reddit, Digg and Epinions; however, the mean-field nature of it limits its calibration with real-world datasets. Medvedev et al. [39] used a Hawkes process along with its branching tree interpretation to jointly model structure and dynamics of discussion trees in Reddit. The model was further used for prediction of the discussion flow, performing better than contemporary models of cascade dynamics.

The discussion trees in Reddit exhibit interesting peculiarities in their structure. For instance, the trees exhibit the so-called "root bias", which means that regardless of the tree size the degree of a root is on average larger and more broadly distributed than the degree of a comment. This may arise from the fact that the direct comment is induced by the post itself, not the subsequent discussion, and that such discussion is only shown after opening a specific link under the post. Root bias is not a unique feature of Reddit, other platforms like Slashdot, Barrapunto, Meneame and Wikipedia also own it [19, 20]. Analysis of political discussion trees [21] and reply trees in Twitter [46] suggests considering the width/depth relation for the trees. For example, reply trees in Twitter were shown to have a duality of being either long chain-like trees (low width, large depth) or star-like trees (vice versa). It is a commonly known fact that critical branching trees of size n have depth d and width w proportional to $\sqrt{n}$ [38], which turns out to be the case for the Reddit discussion trees, where the scaled value of $d/\sqrt{n}$ is well-centred (see Fig. 5). This

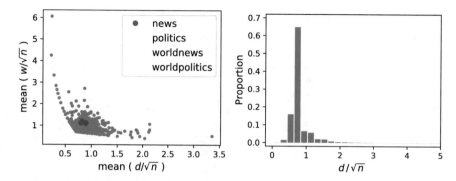

Fig. 5 Average scaled discussion tree depth $d/\sqrt{n}$ versus discussion tree width $w/\sqrt{n}$, where n denotes scaling by tree size (left figure) and scaled depth distribution (right figure) in Reddit. Large red dots on the left figure denote political and news subreddit. The figures are based on data from year 2008

is also supported by the fact that the branching tree model reproduced the trees better than the PA model [39], which apparently has not more than logarithmic depth [9, 12]. Political discussion trees on Slashdot were shown to have larger depth and width in comparison to other trees. Discussions in political subreddits also show large depth and width, but apparently this is mainly due to the fact that on average it has larger participation rate, which is clear from the mean scaled width and depth values for different subreddits in Fig. 5.

Other Zannettou et al. [62] perform a study of meme evolution and propagation across different platforms, e.g. Twitter, Reddit, 4chan and Gab, where the last two are image boards structurally similar to Reddit. The authors' analysis helps to reveal the influential actors in meme ecosystem, both in terms of creation and propagation, the authors build clusters of similar memes and make an analysis of reciprocal influence between the observed communities using Hawkes processes.

The results highlighted in this section are summarized in Table 1.

4 From the Perspective of Users

So far we emphasized the posts as the central pieces of information driving the dynamics of the platform. We now focus on the person who hides behind each post, comment, like or dislike, and review studies on the behavioural features of users. The main approaches in this section are observational and data-oriented. Statistical methods are employed in order to analyse the users track records and their community organization is uncovered by analyzing the network of relations between actors.

Table 1 Short summary of the articles with studies on Reddit, presented in Sect. 3

Article	Task	Dataset	Methods
Horne et al. [29]	Predict high-scoring comments, assess the impact of thread moderation	Reddit dataset [54], 11 top subreddits	Linear regression, sentiment analysis
Fang et al. [13]	Predict final score of comments	Reddit, three chosen subreddits	Recurrent neural networks (RNN)
Zayats and Ostendorf [63]	Predict final score of comments	Reddit, three chosen subreddits	RNN with long short term memory (LSTM)
Hessel et al. [26]	Given a pair of submissions, predict the one with higher final score	Reddit dataset [54], six image-sharing subreddits	Image description (convolutional neural networks), LSTM
Stoddard [52]	Determine inherent quality of posts and to predict high-scoring posts	Hacker News; Reddit dataset [54], five top subreddits	Poisson processes
Lakkaraju et al. [34]	Predict popularity of resubmitted content	Reddit, unique dataset of resubmitted images	Poisson regression
Aragón et al. [3]	Review of the models of discussion trees	Reddit, Slashdot, Meneame, Barrapunto, etc.	Review
Medvedev et al. [39]	Model structure and predict dynamics of discussion trees	Reddit, dataset [54]	Stochastic Hawkes processes

Activity Patterns Observing the actions of users on a website can lead to interesting conclusions. Glenski et al. [16, 17] have studied a dataset with all recorded activity of 309 Reddit users within 1 year. The activity log included the information on all clicks, pageloads and votes made within the REDDIT.COM domain. As expected, the majority of users prefer passive browsing and rarely interact with the content (only 16% of users produce more than 50% of interactions). Users mostly vote on posts on average after only browsing the title (73% of posts), although a non-negligible fraction (17%) of participants follow the link of the post and browse the section of comments before giving a vote. It was noted that users' probability of interaction with a given post decreases with the ranking of the post on the top page of Reddit as well as on subreddits. Text analysis of post titles shows that the probability of interaction increases with the reading ease of the title, i.e. as they use shorter words, and smaller sentences. The authors used the concept of activity sessions, which are the periods of user activity starting from an interaction and finishing after 1 h without consecutive interactions. This terminology and a 1 h threshold were adopted from Halfhaker et al. [22] and Singer et al. [51]. The authors reported a mean session length of 53 min. However, most participants

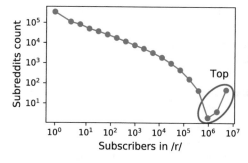

Top subreddits		
AskReddit	movies	Jokes
funny	Music	food
pics	aww	gadgets
worldnews	news	history
videos	books	philosophy
gaming	television	technology
blog	sports	bestof

Fig. 6 Distribution of the number of subscribers of subreddits. The largest subreddits in the rising tail of this distribution are shown along in the table

had much shorter sessions prevailing (>3 min). Singer et al. [51] also studied performance deterioration within sessions of active commenting on Reddit, where sessions of increasing intensity, i.e. how many posts users produced during sessions, are associated with the production of shorter, progressively less complex comments, which receive declining score. In this work, the authors found a similar prevalence of short sessions and sessions and presented a daily circadian rhythms.

Community Loyalty While some of the previous results also applied to other online platforms, this section is devoted specifically to Reddit, a platform where the content is by default submitted to thematic communities, or subreddits. These subreddits are in principle open to anyone, but users can follow particular subreddits of their interest in order to customize their feed. Interestingly, subreddit sizes are close to have heavy-tailed distribution and there exists a fraction of subreddits-outliers with a huge number of subscribers, Fig. 6. Looking closer at those we find that many of these "top" subreddits are those which are proposed to sign up for by default at the registration of a new user.

Tan and Lee [58] studied the posts of users across the subreddits and found that users on average tend to explore and continuously post in new communities; moreover, they tend over time to share their activity evenly between a small number of communities with diverse interests. Differences in posting patterns of users may be used for prediction of the users' future settlement status in a community. Vagrant users, on average, post to more similar communities in comparison with the settling users, they use different language patterns from those existing in a community and their posts receive less attention in terms of score. Score of the first post may generally act as a predictor of further postings. The posting activity rate alone showed to be a bad predictor of the future settlement status in a community. An interesting finding is that the very same users tend to use different vocabulary when posting in different communities, therefore adapting to the community language. Hamilton et al. [23] defined loyal communities as the ones that retain their loyal users over time and find that such communities have smaller, but denser user interaction networks—with users as nodes, connected if there is a reply-to comment between them. These networks were found to be less assortative and less clustered, and thus show less fragmentation into groups. The authors then predicted whether a user will be loyal to the community, using a machine learning classifier with linguistic features of user's posts and comments, and achieve on average 63.6% classification accuracy.

Reddit allows users to self-organize into interest communities, which leads to interesting dynamics of communities. Hessel et al. [25] focused on communities sharing name affixes, for example, affix "ask" (science, askscience), "true" (atheism, trueatheism), "help" (tech, techhelp), etc. A curious finding is that when such a highly related community is created, users engaging in the newer community tend to be more active than in their original one. However, in a prevailing number of cases, newer related subreddits do not detach out of their old partners (in terms of the user base), but in about 25% of the cases, the newer subreddit overtakes its counterpart in participation rate. Some reasons for this behaviour may include the absence of moderation in the new community, a more general scope, or simply a more appealing name spelling. The authors note an interesting result that users who explore the newer created communities generally become relatively more active in their home communities instead of being distracted (Table 2).

Users may also migrate under external pressure. In 2015, a series of external events triggered closure of several popular subreddits. Newell et al. [45] studied the history of this unrest period and observed that users migrating to new subreddits increase their level of participation with respect to their previous community. The authors followed users when they migrated to other discussion platforms and observed that although alternative platforms deliver a space for a broad audience, Reddit users value its advantage of hosting niche communities. In a similar vein, Zhang et al. [64] created a scalable framework for typing subreddits along the "niche" and "volatile" dimensions and used these types to understand the user retention and assimilation in subreddits. Finally, Muchnik et al. [44] performed a large-scale experiment on a Reddit-like platform to study the herding effect of social influence and how the system reacts to the manipulation of comment scores. They observed that users tend to correct artificially downvoted comments. However, comments that are artificially upvoted received an enhanced number of positive votes, thereby increasing the initial bias. Similar herding effects were found in other social systems as well [24, 48]. Das and Lavoie [10] also used a self-collected Reddit dataset of users posts and comments to train a reinforcement-learning model for how users select subreddits to post in reaction to community feedback.

Trolling and Hate Speech Chandrasekharan et al. [7] studied the ban of several hate speech subreddits and the consequences that this measure brought to the website. As one could expect the users of the banned subreddits would redistribute themselves over other subreddits and proceed producing hate speech there, but the authors show that did not happen—the level of hate speech did not increase in other subreddits and, moreover, the majority of users just left the website. Same topic was studied in [47], where the authors note that many counter-actions taken by the users of banned subreddits were short-lived and promptly neutralized by both Reddit moderators.

Table 2 Short summary of the articles with studies on Reddit, presented in Sect. 4

Article	Task	Dataset	Methods
Glenski et al. [16, 17]	Collect and assess the dataset of tracks of user actions	Reddit, unique dataset of user interactions	Statistical analysis
Singer et al. [51]	Assess user performance deterioration during activity sessions	Reddit, all comments made in April 2015	Statistical analysis, negative binomial and Poisson regression
Tan and Lee [58]	Study explorers and exploring phenomena of new communities	Reddit, dataset [54]	Statistical analysis, regression, linear classification
Hamilton et al. [23]	Loyalty prediction for newcoming users, patterns of loyal communities	Reddit, all comments made in 2014	User interaction networks, random forest classifiers
Hessel et al. [25]	Study the dynamic of arise of highly related communities	Reddit, dataset [54]	Statistical analysis
Newell et al. [45]	Study the user migration across platforms during externally caused unrest period	Reddit, dataset [54]	Statistical analysis
Zhang et al. [64]	Classify subreddits along "niche" and "volatile" dimensions, study user retention	Reddit, dataset [54]	Statistical analysis
Das and Lavoie [10]	Model users posting strategies with respect to community feedback	Self-collected Reddit dataset	Machine learning, reinforcement learning, hierarchical Dirichlet process
Kumar et al. [33]	Mobilization and attacks between communities	Reddit, dataset [54]	Reply networks, lexical analysis, LSTM, mechanical turk
Tan [57]	Genealogy of subreddits	Reddit, dataset [54]	Relational networks

In [8] Chandrasekharan et al. studied community norms and their violations. The method comprised continuous scraping of comments and checking their presence in the system 24 h later, which finally gave more than four millions comments deleted by moderators within a 10-month period. Using state-of-the-art text analysis libraries and principle component analysis the common and particular language norms were inferred, for example, hate speech, racism and homophobia were established to be common norms across the whole Reddit and expressing "thanks", mocking religion and nationality were only particular norms, valuable only for a part of subreddits.

Another phenomenon which frequently happens in online discussions is *trolling*—provocative, offensive or menacing messaging [6]. Mojica [41] collected and studied an annotated dataset of trolling comments in discussions on Reddit using a variety of language features.

Inherent Networks of Communities Kumar et al. [33] considered interactions between communities in the form of mobilization by users of a community (the source of the "attack") for hateful comments on posts from another community (the target of the attack). Such mobilization happens when a user in source community posts a link to a post in a target community and titles it with the intention of mobilizing a subset of users, who further write hateful comments on the target post. Such interactions may cause users of a target community to leave. By analyzing reply networks in target discussions, the authors found the effect of echo chambers, i.e. attackers preferentially interact with other attackers and defenders with other defenders. When a direct interaction happens, the attackers "gang-up" on defenders and only a small part of the defenders is involved into interactions with the attackers. The authors propose an LSTM neural network model that uses textual and social features in order to identify whether a given cross-linked post will produce a mobilization.

Gomez et al. [18] constructed and analysed the inherent social network of Slashdot, generated by replies in the discussion threads. The network exhibits neutral mixing by degree, almost identical in and out degree distributions, only moderated reciprocity and an absence of a community structure. The authors conjectured that users are more inclined to be linked to people who express different points of view, and that the network may help to identify users with a high diversity in opinions. The authors also proposed a measure for the controversy of a discussion, based on the h-index [27].

Tan [57] considered the genealogy of communities in Reddit. The author builds a weighted directed network of communities, where community A is linked to a community B if a substantial fraction of first 100 posting users in B had their posts in A. The weight of a link (A, B) is simply the fraction of posting users. The network shows user migration across the communities and is useful for predicting growth of communities. One finds that the diverse portfolio of memberships is the most important characteristic of early adopters, whereas community feedback and language similarity does not seem to matter.

Other Discussion platforms are tested for a broad spectrum of possible research questions. In addition to the topics covered above, we give now some other research directions. Derczynski and Rowe [11] used Reddit comments to create an annotated corpus of *named entities*—proper nouns representing a person, place or an organization. Horne and Adali [28] studied how posting news articles on subreddit */r/worldnews* influences their popularity and concluded that changing the article titles results in greater popularity comparing to leaving the original one. In a similar way, Moyer et al. [43] studied how posts on the subreddit */r/todayilearned* influenced the pageviews of Wikipedia.

The results highlighted in this section are summarized in Table 2.

5 Discussion

This survey does not aim at providing a comprehensive listing of all Reddit-related works, but rather at providing a representative sampling that illustrates the richness of datasets related to online discussion platforms, with Reddit as a dominant example. The richness is clear in terms of quantity as well as quality—in principle the entire dataset can be harvested, while other platforms, e.g. Facebook, may offer exhaustivity only at the cost of selecting a small sample of volunteers [35] and studies of Twitter are known to be limited by the volume and bias of their API [42]. The richness also arises from the diversity of the data, featuring an inherent social network between users, texts constitutive of posts and comments, social appreciation (score), tree structure of posts and comments, all that unfolding in time for years.

Different platforms may exhibit different type of properties, for instance, in the structure of discussion trees, even if they are organized by the same principle. Our discussion on Reddit has shown a broad distribution in size, with a vast majority of trees negligibly small. We also observed root bias, as in other discussion platforms; however, their structure was better modelled as a branching tree with almost uniform branching, rather than preferential attachment networks, due to their specific depth/width profile.

The richness of the data translates in a variety of topics of investigation. On the theoretical level, it allows to observe an ecosystem of users discussing, agreeing or not, and organizing in communities. Besides fundamental sociological questions, it also allows to investigate a range of Internet-specific questions, such as trolling, echo chambers, polarization, social manipulation, etc. The data also offers material to shape solutions for applied questions. From a platform designer viewpoint, it could help to improve the experience of a user, but also to design more efficient algorithms for the identification of high-quality posts. The design of the commenting system is also expected to affect the dynamics and structure of conversations. In this direction, important problems include the detection and automatic removal of trolling or attacks, as well as ways to stimulate the activity of a forum.

The richness of the data and problems calls for a range of computational methods, which may be explicit statistical models or black-box machine learning tools, in order to classify or predict the behaviours of users, posts and communities. Overall we observe that the structure of discussion trees is relatively well understood. However, mixing the dynamics and structure with textual features is an important step that has only been studied by means of black-box machine learning, such as neural networks techniques, showing a good performance in predicting community appreciation. A challenge remains the fact that many basic statistics of the data (activity of users, popularity of communities, success of a post, etc.) exhibit heavy tails, which may introduce sampling issues as, for instance, a random sampling may fail to observe extreme points (e.g. high activity users) while they carry a large influence in the structure and dynamics of the system. This caveat must be kept in mind when using techniques such as neural networks.

This review is a testimony of the richness and dynamism of academic research on social platforms, in general, and Reddit, in particular. Despite the many progresses overviewed above, we would like to conclude with a list of what we believe to be promising research directions. In our opinion, fruitful avenues of research include:

- A more detailed study of the activity patterns of users. Collected browsing patterns of users already uncover particular voting behaviour, when many users vote after glancing over the title of a post, and browsing patterns, when feed page breaks create an abrupt obstacle for users attention [16, 17]. Activity patterns may first of all be of use to platform designers and to study the influence of platform structure on user experience.
- It is true that discussion platforms like Reddit do not have an a priori built social network, in comparison with Twitter or Facebook. Nevertheless, one may reconstruct inherent networks of communities [57], users (as in [18]) or submitted information, and exploit these networks to improve prediction of the future state or dynamics of the system. Such studies may be of use for platform users, as well as for platform curators.
- Study growth or resilience of online communities over time. To be more precise, the dynamics of posting and commenting shape define the life of a community. The two processes are coupled, but not necessarily proportional, as can be seen in Fig. 7. Important questions include the identification of dynamical and structural features that ensure the growth or resilience of online communities over time. Dynamical and evolving graph models would be of help in this direction.
- The dynamics of discussions is another interesting, yet mostly unexplored, aspect of research, especially the possible relation between the structure and the dynamics of discussion trees. This question could be explored by means of neural networks, which showed to be a working approach in prediction models.
- Finally, the huge volume of new posts and comments makes the design of efficient ranking and recommendation algorithms vital, in order to allow users to identify relevant information and improve their online experience. As it was shown, platform structure has direct influence on user experience and participation [3]; thus, both platform designers and users would benefit from such studies.

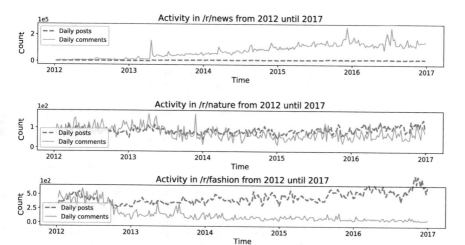

Fig. 7 Daily counts of submission of posts and comments in three selected subreddits. Three possible scenarios of participation dynamics are shown: (1) increase in comment rate exceeds posting rate (top figure); (2) rates are similar (middle figure); and (3) comments eventually disappear, while the number of posts increases

Acknowledgements This work was supported by Concerted Research Action (ARC) supported by the Federation Wallonia-Brussels Contract ARC 14/19-060; Flagship European Research Area Network (FLAG-ERA) Joint Transnational Call "FuturICT 2.0"; and by grant 16-01-00499 of the Russian Foundation for Basic Research.

References

1. Aragón, P., Gómez, V., Kaltenbrunner, A.: Visualization tool for collective awareness in a platform of citizen proposals. In: Proceedings of the International AAAI Conference on Weblogs and Social Media, pp. 756–757 (2016)
2. Aragón, P., Gómez, V., García, D., Kaltenbrunner, A.: Generative models of online discussion threads: state of the art and research challenges. J. Internet Serv. Appl. **8**(1), 15 (2017)
3. Aragón, P., Gómez, V., Kaltenbrunner, A.: To thread or not to thread: the impact of conversation threading on online discussion. In: International AAAI Conference on Web and Social Media (2017)
4. Backstrom, L., Boldi, P., Rosa, M., Ugander, J., Vigna, S.: Four degrees of separation. In: Proceedings of the 4th Annual ACM Web Science Conference, pp. 33–42. ACM, New York (2012)
5. Bandari, R., Asur, S., Huberman, B.A.: The pulse of news in social media: forecasting popularity. In: Proceedings of the Sixth International AAAI Conference on Weblogs and Social Media (ICWSM), vol. 12, pp. 26–33 (2012)
6. Bishop, J.: The effect of de-individuation of the internet troller on criminal procedure implementation: an interview with a hater. Int. J. Cyber Criminol. **7**(1), 28–48 (2013)
7. Chandrasekharan, E., Pavalanathan, U., Srinivasan, A., Glynn, A., Eisenstein, J., Gilbert, E.: You can't stay here: the efficacy of Reddit's 2015 ban examined through hate speech. Proc. ACM Hum.-Comput. Interact. **1**, 31 (2017)

8. Chandrasekharan, E., Samory, M., Jhaver, S., Charvat, H., Bruckman, A., Lampe, C., Eisenstein, J., Gilbert, E.: The internet's hidden rules: an empirical study of Reddit norm violations at micro, meso, and macro scales. Proc. ACM Hum.-Comput. Interact. **2**, 32:1–32:25 (2018). http://doi.acm.org/10.1145/3274301

9. Cohen, R., Havlin, S.: Scale-free networks are ultrasmall. Phys. Rev. Lett. **90**(5), 058701 (2003)

10. Das, S., Lavoie, A.: The effects of feedback on human behavior in social media: an inverse reinforcement learning model. In: Proceedings of the 2014 International Conference on Autonomous Agents and Multi-Agent Systems, pp. 653–660. International Foundation for Autonomous Agents and Multiagent Systems (2014)

11. Derczynski, L., Rowe, M.: Tracking the diffusion of named entities. (2017, preprint). arXiv:1712.08349

12. Dommers, S., Van Der Hofstad, R., Hooghiemstra, G.: Diameters in preferential attachment models. J. Stat. Phys. **139**(1), 72–107 (2010)

13. Fang, H., Cheng, H., Ostendorf, M.: Learning latent local conversation modes for predicting comment endorsement in online discussions. In: Proceedings of The Fourth International Workshop on Natural Language Processing for Social Media, pp. 55–64 (2016)

14. Gaffney, D., Matias, J.N.: Caveat emptor, computational social science: large-scale missing data in a widely-published Reddit corpus. (2018, preprint). arXiv:1803.05046

15. Gilbert, E.: Widespread underprovision on Reddit. In: Proceedings of the 2013 Conference on Computer Supported Cooperative Work, pp. 803–808. ACM, New York (2013)

16. Glenski, M., Weninger, T.: Predicting user-interactions on Reddit. In: Proceedings of the 2017 IEEE/ACM International Conference on Advances in Social Networks Analysis and Mining 2017, pp. 609–612. ACM, New York (2017)

17. Glenski, M., Pennycuff, C., Weninger, T.: Consumers and curators: browsing and voting patterns on Reddit. IEEE Trans. Comput. Soc. Syst. **4**(4), 196–206 (2017)

18. Gómez, V., Kaltenbrunner, A., López, V.: Statistical analysis of the social network and discussion threads in Slashdot. In: Proceedings of the 17th International Conference on World Wide Web, pp. 645–654. ACM, New York (2008)

19. Gómez, V., Kappen, H.J., Kaltenbrunner, A.: Modeling the structure and evolution of discussion cascades. In: Proceedings of the 22Nd ACM Conference on Hypertext and Hypermedia, pp. 181–190 (2011)

20. Gómez, V., Kappen, H.J., Litvak, N., Kaltenbrunner, A.: A likelihood-based framework for the analysis of discussion threads. World Wide Web **16**(5–6), 645–675 (2013)

21. Gonzalez-Bailon, S., Kaltenbrunner, A., Banchs, R.E.: The structure of political discussion networks: a model for the analysis of online deliberation. J. Inf. Technol. **25**(2), 230–243 (2010). https://doi.org/10.1057/jit.2010.2

22. Halfaker, A., Keyes, O., Kluver, D., Thebault-Spieker, J., Nguyen, T., Shores, K., Uduwage, A., Warncke-Wang, M.: User session identification based on strong regularities in inter-activity time. In: Proceedings of the 24th International Conference on World Wide Web, pp. 410–418. International World Wide Web Conferences Steering Committee, Geneva (2015)

23. Hamilton, W.L., Zhang, J., Danescu-Niculescu-Mizil, C., Jurafsky, D., Leskovec, J.: Loyalty in online communities. In: Proceedings of the International AAAI Conference on Weblogs and Social Media, vol. 2017, p. 540. NIH Public Access (2017)

24. Hanson, W.A., Putler, D.S.: Hits and misses: herd behavior and online product popularity. Mark. Lett. **7**(4), 297–305 (1996)

25. Hessel, J., Tan, C., Lee, L.: Science, askscience, and badscience: on the coexistence of highly related communities. In: The Tenth International Conference on Web and Social Media (ICWSM), pp. 171–180 (2016)

26. Hessel, J., Lee, L., Mimno, D.: Cats and captions vs. creators and the clock: comparing multimodal content to context in predicting relative popularity. In: Proceedings of the 26th International Conference on World Wide Web, pp. 927–936. International World Wide Web Conferences Steering Committee, Geneva (2017)

27. Hirsch, J.E.: An index to quantify an individual's scientific research output. Proceedings of the National Academy of Sciences **102**(46), 16569–16572 (2005)

28. Horne, B.D., Adali, S.: The impact of crowds on news engagement: a Reddit case study. (2017, preprint). arXiv:1703.10570
29. Horne, B.D., Adali, S., Sikdar, S.: Identifying the social signals that drive online discussions: a case study of Reddit communities. In: 26th International Conference on Computer Communication and Networks (ICCCN), pp. 1–9 (2017). https://doi.org/10.1109/ICCCN.2017.8038388
30. Jaech, A., Zayats, V., Fang, H., Ostendorf, M., Hajishirzi, H.: Talking to the crowd: what do people react to in online discussions? In: Proceedings of the 2015 Conference on Empirical Methods in Natural Language Processing, pp. 2026–2031 (2015)
31. Kaltenbrunner, A., Gomez, V., Lopez, V.: Description and prediction of Slashdot activity. In: Latin American Web Conference 2007 (LA-WEB 2007), pp. 57–66. IEEE, Piscataway (2007)
32. Karsai, M., Kivelä, M., Pan, R.K., Kaski, K., Kertész, J., Barabási, A.L., Saramäki, J.: Small but slow world: how network topology and burstiness slow down spreading. Phys. Rev. E 83(2), 025102 (2011)
33. Kumar, S., Hamilton, W.L., Leskovec, J., Jurafsky, D.: Community interaction and conflict on the web. In: Proceedings of the 2018 World Wide Web Conference on World Wide Web, pp. 933–943. International World Wide Web Conferences Steering Committee, Geneva (2018)
34. Lakkaraju, H., McAuley, J.J., Leskovec, J.: What's in a name? Understanding the interplay between titles, content, and communities in social media. In: International AAAI Conference on Web and Social Media (ICWSM), vol. 1, no. 2, 3 (2013)
35. Lambiotte, R., Kosinski, M.: Tracking the digital footprints of personality. Proc. IEEE 102(12), 1934–1939 (2014)
36. Lee, J.G., Moon, S., Salamatian, K.: Modeling and predicting the popularity of online contents with cox proportional hazard regression model. Neurocomputing 76(1), 134–145 (2012)
37. Lumbreras, A., Jouve, B., Velcin, J., Guégan, M.: Role detection in online forums based on growth models for trees. Soc. Netw. Anal. Min. 7(1), 49 (2017)
38. Marckert, J.F., Mokkadem, A., et al.: The depth first processes of Galton–Watson trees converge to the same Brownian excursion. Ann. Probab. 31(3), 1655–1678 (2003)
39. Medvedev, A.N., Delvenne, J.C., Lambiotte, R.: Modelling structure and predicting dynamics of discussion threads in online boards. J. Complex Netw. 7, 67–82 (2018). https://doi.org/10.1093/comnet/cny010
40. Mishne, G., Glance, N.: Leave a reply: an analysis of weblog comments. In: Proceedings of 3rd Annual Workshop on the Weblogging Ecosystem at the 15th International World Wide Web Conference (2006)
41. Mojica, L.G.: Modeling trolling in social media conversations. (2016, preprint). arXiv:1612.05310
42. Morstatter, F., Pfeffer, J., Liu, H., Carley, K.M.: Is the sample good enough? comparing data from Twitter's streaming API with Twitter's Firehose. In: International AAAI Conference on Web and Social Media (ICWSM) (2013)
43. Moyer, D., Carson, S.L., Dye, T.K., Carson, R.T., Goldbaum, D.: Determining the influence of Reddit posts on Wikipedia pageviews. In: Proceedings of the Ninth International AAAI Conference on Web and Social Media (2015)
44. Muchnik, L., Aral, S., Taylor, S.J.: Social influence bias: a randomized experiment. Science 341(6146), 647–651 (2013)
45. Newell, E., Jurgens, D., Saleem, H.M., Vala, H., Sassine, J., Armstrong, C., Ruths, D.: User migration in online social networks: a case study on Reddit during a period of community unrest. In: International AAAI Conference on Web and Social Media (ICWSM), pp. 279–288 (2016)
46. Nishi, R., Takaguchi, T., Oka, K., Maehara, T., Toyoda, M., Kawarabayashi, K.I., Masuda, N.: Reply trees in twitter: data analysis and branching process models. Soc. Netw. Anal. Min. 6(1), 1–13 (2016)
47. Saleem, H.M., Ruths, D.: The aftermath of disbanding an online hateful community (2018). Preprint. arXiv:1804.07354
48. Salganik, M.J., Watts, D.J.: Leading the herd astray: an experimental study of self-fulfilling prophecies in an artificial cultural market. Soc. Psychol. Quart. 71(4), 338–355 (2008)

49. Sinatra, R., Lambiotte, R.: Topical issue-quantifying success. Adv. Complex Syst. **21**, 3–4 (2018)
50. Singer, P., Flöck, F., Meinhart, C., Zeitfogel, E., Strohmaier, M.: Evolution of Reddit: from the front page of the internet to a self-referential community? In: Proceedings of the 23rd International Conference on World Wide Web, pp. 517–522. ACM, New York (2014)
51. Singer, P., Ferrara, E., Kooti, F., Strohmaier, M., Lerman, K.: Evidence of online performance deterioration in user sessions on Reddit. PloS One **11**(8), e0161636 (2016)
52. Stoddard, G.: Popularity dynamics and intrinsic quality in Reddit and hacker news. In: International AAAI Conference on Web and Social Media (ICWSM), pp. 416–425 (2015)
53. Stuck_In_the_Matrix: Dataset is available on the following webpage. https://files.pushshift.io/reddit/ (Query: 2017-06-01)
54. Stuck_In_the_Matrix: I have every publicly available Reddit comment for research. approx. 1.7 billion comments @ 250 gb compressed. any interest in this? https://redd.it/3bxlg7 (Query: 2017-07-14)
55. Stuck_In_the_Matrix: Update for the Reddit corpus. https://redd.it/8aen5g (Query: 2018-09-27)
56. Szabo, G., Huberman, B.A.: Predicting the popularity of online content. Commun. ACM **53**(8), 80–88 (2010)
57. Tan, C.: Tracing community genealogy: how new communities emerge from the old. (2018, preprint). arXiv:1804.01990
58. Tan, C., Lee, L.: All who wander: on the prevalence and characteristics of multi-community engagement. In: Proceedings of the 24th International Conference on World Wide Web, WWW '15, pp. 1056–1066. International World Wide Web Conferences Steering Committee, Republic and Canton of Geneva (2015)
59. Tsagkias, M., Weerkamp, W., De Rijke, M.: Predicting the volume of comments on online news stories. In: Proceedings of the 18th ACM Conference on Information and Knowledge Management, pp. 1765–1768. ACM, New York (2009)
60. Wakefield, J.: Are you scared yet? Meet Norman, the psychopathic AI. BBC News https://www.bbc.com/news/technology-44040008
61. Wang, C., Ye, M., Huberman, B.A.: From user comments to on-line conversations. In: Proceedings of the 18th ACM SIGKDD International Conference on Knowledge Discovery and Data Mining, KDD'12, pp. 244–252 (2012)
62. Zannettou, S., Caulfield, T., Blackburn, J., De Cristofaro, E., Sirivianos, M., Stringhini, G., Suarez-Tangil, G.: On the origins of memes by means of fringe web communities (2018). Preprint. arXiv:1805.12512
63. Zayats, V., Ostendorf, M.: Conversation modeling on Reddit using a graph-structured LSTM. Trans. Assoc. Comput. Linguist. **6**, 121–132 (2018)
64. Zhang, J., Hamilton, W.L., Danescu-Niculescu-Mizil, C., Jurafsky, D., Leskovec, J.: Community identity and user engagement in a multi-community landscape. In: Proceedings of the International AAAI Conference on Weblogs and Social Media, vol. 2017, p. 377. NIH Public Access (2017)
65. Zhao, Q., Erdogdu, M.A., He, H.Y., Rajaraman, A., Leskovec, J.: Seismic: A self-exciting point process model for predicting tweet popularity. In: Proceedings of the 21th ACM SIGKDD International Conference on Knowledge Discovery and Data Mining, pp. 1513–1522 (2015)

Learning Information Dynamics in Online Social Media: A Temporal Point Process Perspective

Bidisha Samanta, Avirup Saha, Niloy Ganguly, Sourangshu Bhattacharya, and Abir De

Abstract Accurate modeling of information dynamics of online media across time has a wide variety of applications. For example, in Twitter, we can predict which hashtag may go viral against others; also in an e-commerce site like Amazon, reviews of a product over the time can help to identify which product will be preferred over others in future. The information dynamics follow a complex diffusion process and many factors reinforce each other. There are clearly two types of factors: (a) intra-item factors and (b) inter-item factors. Visibility indicates the ability of a piece of information to attract the attention of the users, against the background information. Therefore, apart from the individual information diffusion processes, the information visibility dynamics also involves a competition process, where each information diffusion process competes against others to draw the attention of users. Despite the fact that models of the individual information diffusion process abound in literature, modeling the competition process is left unaddressed. Here, we propose two models: (1) LMPP: a probabilistic linear framework that unifies influence of different factors contributing to popularity of an item and inter-item (product or hashtag) competitions and (2) CRPP: a more generic model, a probabilistic deep machinery that unifies the nonlinear generative dynamics of a collection of diffusion processes, and inter-process competition— the two ingredients of visibility dynamics. Though LMPP is a novel probabilistic lightweight framework that models the dynamics of item popularity by unifying the intra- and inter-influence in a principled way, it assumes a linear diffusion process which makes it restrictive in some cases. CRPP overcomes this issue. To design this model, we rely on a recurrent neural network (RNN) guided generative framework, where the recurrent unit captures the joint temporal dynamics of a group

B. Samanta · A. Saha (✉) · N. Ganguly · S. Bhattacharya
Indian Institute of Technology Kharagpur, Kharagpur, India
e-mail: bidisha@iitkgp.ac.in; avirupsaha@iitkgp.ac.in; niloy@cse.iitkgp.ernet.in; sourangshu@cse.iitkgp.ernet.in

A. De
Max Planck Institute for Software Systems, Saarbrücken, Germany
e-mail: ade@mpi-sws.org

© Springer Nature Switzerland AG 2019
F. Ghanbarnejad et al. (eds.), *Dynamics On and Of Complex Networks III*,
Springer Proceedings in Complexity, https://doi.org/10.1007/978-3-030-14683-2_10

of processes. This is aided by a discriminative model which captures the underlying competition process by discriminating among the various processes using several ranking functions. On ten diverse datasets crawled from Amazon and Twitter, CRPP and LMPP offer a substantial performance boost in predicting product visibility against several baselines, thereby achieving significant accuracy in predicting both the collective diffusion mechanism and the underlying competition processes.

Keywords Information diffusion · Social network · Machine learning

1 Introduction

With the explosion of information spread in several online forums, fuelled by the enormous activities by their users, the user-attention has become a scarce yet valuable commodity [4]. Consequently, among a myriad of information, only a few can steal a good part of it, while the rest go unnoticed. For example, in Twitter, a small number of hashtags attract a frenetic level of retweets, while the others quickly ebb; in e-commerce sites like Amazon, Alibaba, etc., only a few popular products are swamped with reviews, and the rest remain confined in small coteries. Such a trait is called *visibility* which refers to the relative ability of a particular piece of information to draw the attention of the users, against the rest. We quantify visibility of a piece of information by the number of instances where it is cited by users, e.g., by the number of (re)tweets bearing a hashtag. In general, the popularity of a piece of information depends on one more factor other than visibility and competition— that is intra-item reinforcement. Every hashtag has an intrinsic attractiveness, and similarly the tweets bearing the hashtag also have their own appeal. In one of our works [27], we showed that hashtags and tweets often reinforce each other. For example a not-so-popular tweet may become popular later on due to the popularity of the hashtag it is bearing. Since traditional temporal approaches bank on modeling only tweet-propagation, simply extending these prediction-frameworks to hashtags would not produce accurate results (our experimental results also emphatically establish that). However, considering only the hashtag-tweet reinforcement process still leaves a paucity in the realistic modeling of hashtag-flow, that demands a careful consideration of the inter-hashtag competitions.

 Therefore, at the very outset, the information dynamics that involves a complex, collective information diffusion process is also propelled by a competition mechanism among several pieces of information, where one is pitted against the others for drawing a significant share of user attention [14, 29, 33]. Principled modeling of information visibility can potentially have immense impact, on a broad application spectrum ranging from trending hashtag selection to news broadcasting, from information spread to product purchase prediction, etc. In this paper, our goal is to accurately model such visibility dynamics by unifying the joint generative diffusion mechanism of a group of information, and the competitive interaction between them, in a principled way.

Limitations of Prior Work

Research on online information diffusion that was based on the discrete time cascade models during the last decade [2, 3, 15, 16, 24] recently got a fresh lease with the advent of temporal point process—a stochastic process—which naturally captures the asynchronous arrival of information in online media [1, 6, 8, 10, 17, 22, 23, 30, 34]. While such techniques have shown a great finesse in predicting the popularity of individual information, their "as-it-is" deployment to model the visibility dynamics is precluded mainly by two limitations. (1) They do not explicitly capture the competition between two active pieces of information. Such a competition between information pieces is also a complex dynamical process, where the popularity rankings of the competing information pieces continuously fluctuate with time. (2) The existing works focus on generative modeling for individual information flows. By doing so, they skirt several fine-grained traits involving competing dynamics of information, e.g., relative popularity variation, sudden trend change, etc. Consequently, they are largely unable to replicate any microscopic features in information diffusion. While a few works [30] indeed coin the issue of competition in the context of product adoption, they adopt a brittle linear model for information diffusion that renders their approach practically restrictive. Our experimental results establish this fact (Sect. 3).

Present Work

We first describe **L**arge **M**argin **P**oint **P**rocess (LMPP), a novel probabilistic framework that models the information dynamics by unifying the intra-item factor and inter-item competition. LMPP aims to capture item-process reinforcement using a generalized triggering kernel. Furthermore, LMPP aims to incorporate competition among the items and their impact on information dynamics. In order to do that, we probe into the variations in the popularity rankings of the concurrent and related items. Therefore, to capture such signals, we suitably curate the parameter space of LMPP that ensures correct ordering of popularity across several time intervals. Such a formulation intuitively articulates the competition process, without drastically changing the model-setting. In fact, this additional trait helps to properly train the model, which in turn enables it to detect sudden drifts in popularity rankings of the competing items.

However, LMPP adopts a linear model for information diffusion that renders their approach practically restrictive. To ameliorate the above limitations, in this paper, we develop **C**ompeting **R**ecurrent **P**oint **P**rocess (CRPP) for modeling information visibility dynamics. CRPP unifies its two key ingredients: (1) a discriminative competition framework to tap the relative popularity variation between several competing diffusion processes, which in turn are modeled using a family of coupled (multivariate) nonlinear temporal point processes; and (2) a multivariate nonlinear generative module to probe the dependent dynamics of these processes. More in detail, CRPP is equipped with a family of learning-to-rank templates that embed the associated events into real valued vectors using multiple deep recurrent neural networks. These learning-to-rank templates are specifically designed to probe the relative variations in the popularity order of the concurrent processes, and feed these signals into suitable ranking losses. Such an approach bridges the rich literature on learning-to-rank [11–13] with temporal point process, where the ranking losses operate over a complex dynamics of interdependent samples and optimizes a diverse set of visibility measures. Moreover, by doing so, our model also captures the complex generative dynamics of concurrent information diffusion processes by means of the multivariate RNN modules, thereby developing the inchoate ideas of univariate point-process modeling [6, 23] into a complete, generic, multivariate design.

On ten diverse datasets, six obtained from Amazon reviews and four crawled from Twitter, our proposal offers substantial (up to ~18%) accuracy gains in predicting visibility dynamics, against several strong competitors. More importantly, by adequately learning the competition process, it can reasonably detect an abrupt visibility change of an item, i.e., a hashtag or a product, a challenging task that many baselines cannot trace. Such a strong predictive power enables our model to learn to rank products on fly, which is a considerably difficult task yet with a wide spectrum of applications. By accurately considering the competition process, it can successfully model the ranking dynamics over time, of the correlated hashtags or products, which none of the existing baselines can even consistently trace. Consequently, it can reasonably forecast the abrupt popularity changes of these items, which, in general, is considered a difficult phenomenon to reproduce.

Contributions

Apart from designing a novel model for visibility dynamics in the context of information propagation, in this paper, we make the following contributions:

1. Both the models employ a wide variety of learning-to-rank templates, where they minimize a set of ranking losses over the variation of popularity order observed in the data. Despite the complex temporal-dependencies between competing information-streams, the presence of such a learning machinery in our approach helps to connect the rich concepts of learning-to-rank methods to temporal point process—which is a key contribution of our paper.
2. In a marked departure from the existing diffusion models having fixed functional forms, CRPP eschews such impositions and rather learns the complexity of the nonlinear visibility dynamics. Furthermore, our model generalizes the basic ideas of deep learning of univariate point processes to design a complete multivariate deep point process framework.

2   Proposed Model

In this section we will describe two models, starting from the linear large margin point process based model LMPP and then we will describe the more generic nonlinear framework CRPP. Throughout the rest of this paper we will use the term item diffusion and process interchangeably.

2.1 Large Margin Point Process (LMPP)

2.1.1 Overview

Terminology We define a **Diffusion Process** as the set of events. In case of Twitter the diffusion process of a hashtag can be defined as the collection of tweet posting events. In case of the scenario of the products at Amazon, a user review corresponds to a product use event, and the diffusion process is nothing but product popularity dynamics.

Given a process $\mathcal{I}$, events $\{e\}$ we define the history before t as:

$$\mathcal{H}_{\mathcal{I}}(t) = \cup_{e \in \mathcal{I}}\{e \text{ and } t(e) < t\},$$

where $t(.)$ defines the time when the corresponding event occurred. Thus, the history of a process until the time t is the set of events generated by the process before the time t.

Computation of Item Popularity In a similar spirit to [34], we measure the popularity of an item by the total number of its constituent events. To do so, we propose LMPP, that models the temporal dynamics of the constituent events in terms of post-rate. While modeling such a post-rate, LMPP combines the role of event-process reinforcement that is, the popular event would influence the dynamics more than a not-so-popular event and item competitions in the overall popularity dynamics.

Basic Generative Process for an Item $\mathcal{I}$ At the outset, we represent the posting times of the events as a point-process model. In particular, given an item $\mathcal{I}$, we define the counting variable as $N_{\mathcal{I}}(t)$, where $N_{\mathcal{I}}(t) \in \{0\} \cup \mathbb{Z}^+$ counts the number of events posted until and excluding time t. Then, we characterize the conditional probability of observing an event in infinitesimal time interval $[t, t + dt)$ as

$$\mathbb{P}(\text{An event triggers in } [t, t + dt)|\mathcal{H}_{\mathcal{I}}(t)) = \lambda_{\mathcal{I}}(t)\, dt \tag{1}$$

$$\text{i.e.,} \mathbb{E}_{dN_{\mathcal{I}}(t) \sim \{0,1\}}[dN_{\mathcal{I}}(t)|\mathcal{H}_{\mathcal{I}}(t)] = \lambda_{\mathcal{I}}(t)dt. \tag{2}$$

Here $dN_{\mathcal{I}}(t)$ indicates the number of events in the infinitesimal time-window $[t, t + dt)$ and $\lambda_{\mathcal{I}}(t)$ stands for the associated hashtag intensities, which further depends on the history $\mathcal{H}_{\mathcal{I}}(t)$.

Based on the above definition of conditional probability, We may quantitatively define the visibility of an item $\mathcal{I}$ in an interval $[t_s, t_f)$ as $N_{\mathcal{I}}[t_s, t_f) = \int_{t_s}^{t_f} \lambda_{\mathcal{I}}(t)dt$ and the overall visibility of the item in the whole time period $[0, T)$ as $N_{\mathcal{I}}(t) = \int_0^t \lambda_{\mathcal{I}}(t)dt$

The functional form of $\lambda_{\mathcal{I}}(t)$ is chosen to capture the phenomenon of interests that possibly encompass process-competitions, self-exciting dynamics, or event-process interactions. In the following, we present a specific characterization of $\lambda_{\mathcal{I}}(t)$ that captures the self-exciting nature of a process dynamics.

Self-Exciting Dynamics To capture the mutual excitation between posting events, we rely on the *Hawkes process* [5, 7]. It is a particular type of functional form used in the growing literature on social activity modeling using point processes [7, 30]:

$$\lambda_{\mathcal{I}}(t) = \lambda_{\mathcal{I},0} + \beta \sum_{t_i \in \mathcal{H}_{\mathcal{I}}(t)} e^{-\omega_0(t-t_i)}$$

$$= \lambda_{\mathcal{I},0} + \beta(\kappa(t) \star dN_{\mathcal{I}}(t)). \tag{3}$$

Here, $\lambda_{\mathcal{I},0} \geq 0$ models the initial post-rate of (re)tweets, and the second term, with $\beta \geq 0$, assigns weight to the influence of the publication of earlier events. $\kappa(t) = e^{-\omega_0 t}$ is an exponential triggering kernel indicating the decay of influence of the past events over time, and $\star$ denotes the convolution operation.

2.1.2 LMPP: Modeling Event-Process Reinforcement and Inter-Process Competitions

Apart from the inherent popularity dynamics, our proposed framework CRPP considers two more crucial factors in modeling the popularity dynamics of an item: (1) the mutual reinforcement process between item and process, and (2) the competitions among processes.

We observe that, it is the intensity kernel $\kappa(t)$ (in Eq. (3)) that accounts for the self-exciting mechanism for the Hawkes process. Therefore, we aim to construct a suitable $\kappa(t)$ which should, along with the self-exciting reinforcement process of the individual tweets, capture the hashtag-tweet reinforcement factor. Therefore, the event-process reinforcement factor should vary across the popularity distribution of the event. Thus, the kernel should be further parameterized by *event-popularity index k* (defined in Sect. 2.1.3), to have $\kappa_k(t)$. Particularly, $\kappa_k(t)$ should be chosen in such a way that:

- Given a process, when a popular event occurs, i.e., when k goes high, the inherent attractiveness of the item heavily influences its propagation process. That is, $\kappa_k(t)$ pushes $\lambda_{\mathcal{I}}(t)$ more towards a Hawkes process, in the sense that the occurrence of an event stimulates a large increase in the probability of events in the immediate future (which then dies down eventually). This is referred to as "self-exciting" dynamics. Therefore, the overall resulting dynamics become more and more bursty, i.e., the distribution of process exhibits several high peaks (and deviates from a uniform distribution). In the case of Twitter, $\kappa_k(t)$ can be looked upon as a normalized measure $(N_i / \sum_j N_j)$ of the number of retweets of the tweet i in the time interval $[0, t]$. Such a term encourages "hashtag-tweet reinforcement," where a popular tweet drives the flow of the underlying hashtag, rather than a not-so-popular tweet. In case of the scenario of the products at Amazon, where a user review corresponds to a product use event, k would denote the rating of the review, which is on a scale of 1–5 stars. $\kappa_k(t)$ can therefore be looked upon as a normalized measure $k/5$. In this case, the event magnitude models what may be analogously called "product-review reinforcement," where a review with a higher star rating boosts the popularity of the product, leading to higher sales and correspondingly more product use events.

- For non-popular events, the effect of the resultant influence of the process is very low on its propagation process. Thus, a low value of $\kappa_k(t)$ should fare in a relatively small $\lambda_{\mathcal{I}}(t)$.

Considering the above points, we take $\kappa_k(t)$ as

$$\kappa_k(t) = \kappa_\infty(t)e^{-\frac{\omega t}{k}}. \tag{4}$$

$\kappa_k(t)$ has two factors. $\kappa_\infty(t)$ indicates the self-exciting process, while $e^{-\frac{\omega t}{k}}$ stands for the event-process influence. Furthermore, we try to approximate $\kappa_\infty(t)$ as a more generalized intensity kernel $\kappa_\infty(t) = \sum_{j=1}^{M} \beta_j e^{-\omega_j t}$, where M is a large integer. The higher the value of M, the more flexible the model. Thus, a high value of M can be used if the available dataset is large and lower values give a poor fit. However, very large values of M can lead to overfitting. Thus, we need to empirically determine the optimum value of M for a dataset. The values used in the experiments are of the order of 10^4.

Then, the arrival rate of tweets can be written as

$$\lambda_{\mathcal{I}}(t; k_t) = \lambda_{\mathcal{I},0}e^{-\epsilon t} + \sum_{j=1}^{M} \beta_{\mathcal{I}}^{j} \sum_{t_i \in \mathcal{H}_{\mathcal{I}}(t)} e^{-(\omega_j + \frac{\omega}{k_{t_i}})(t - t_i)}. \tag{5}$$

Here, $k(t_i)$ is the popularity of an event posted at time t_i, and $k_t := \{k_{t_i} | t_i \in \mathcal{H}_{\mathcal{I}}(t)\}$. For compactness, we denote $\omega = [\omega_1, \omega_2, \ldots \omega_M]$. Here, an additional decay factor $e^{-\epsilon t}$ is incorporated to diminish the effect of the initial condition, which we found to work well in practice. The value of ϵ is chosen empirically. It is the same for all hashtags.

2.1.3 Popularity Distribution

We observe that given a hashtag, the distribution of the popularity indices of individual tweets follows a power-law (figures omitted for brevity), which means that the tweets getting very high retweets are very small in number, whereas plenty of tweets are having small number of retweets. The distribution is captured as below:

$$p(k) = ck^{-\alpha} \text{ with } c = \alpha - 1. \tag{6}$$

where k is the popularity of a tweet-chain.

Hence, the expected arrival rate of the process having tweet-chains with *random popularity* can be formulated as:

$$\tilde{\lambda}_{\mathcal{I}}(t) = \mathbb{E}_k[\lambda_{\mathcal{I}}(t; \boldsymbol{k}(t))] = \int_1^\infty \lambda_{\mathcal{I}}(t; \boldsymbol{k}(t)) ck^{-\alpha} dk, \tag{7}$$

where c is a constant given by Eq. (6).

For product review this can be calculated from the dataset as:

$$\tilde{\lambda}_{\mathcal{I}}(t) = \mathbb{E}_k[\lambda_{\mathcal{I}}(t; \boldsymbol{k}(t))] = \sum_{i=1}^5 \lambda_{\mathcal{I}}(t; \boldsymbol{k}(t)) p(k = i), \tag{8}$$

where $k \in \{1, 2, 3, 4, 5\}$, and assuming a uniform distribution of ratings, $p(k = i) = \frac{1}{5} \forall i = 1(1)5$.

2.1.4 Popularity Ranking in Hashtag Competition

In a diffusion process, concurrent processes often compete with each other for user's attention. Such scenarios are usually pronounced through the variations of popularity rankings of the competing items over time. To model it, one may specify $\lambda_{\mathcal{I}}(t)$, so that it detects the variation in their popularity rankings across time. In particular, we say

$$N_{\mathcal{I}_1}[t_s, t_f) > N_{\mathcal{I}_2}[t_s, t_f) \implies$$
$$\int_{t_s}^{t_f} \lambda_{\mathcal{I}_1}(t)dt \geq \int_{t_s}^{t_f} \lambda_{\mathcal{I}_2}(t)dt + 1 \ \forall \mathcal{I}_1, \mathcal{I}_2 \text{ and } t_s < t_f, \tag{9}$$

where $N_{\mathcal{I}}[t_s, t_f)$ denotes the number of (re)tweets of hashtag $\mathcal{I}$ posted in the interval $[t_s, t_f)$. The "+1" has been included to incorporate a large margin in the constraints. Because the counting process is discrete, the margin must be an integer, and the least value is 1.

2.2 Parameter Estimation

Given a set of n processes $\mathcal{I} = \{\mathcal{I}_l | 1 \leq l \leq n\}$, we record a collection of posts $\mathcal{H}_{\mathcal{I}_l}(T) = \{\tau_i\}$ for each process $\mathcal{I}_l$ during a time period $[0, T)$. Using these posts, we attempt to find the optimal parameters $\lambda_{\mathcal{I},0}$ and $\boldsymbol{\beta}_{\mathcal{I}} = [\beta_{\mathcal{I}}^1, \beta_{\mathcal{I}}^2, \ldots, \beta_{\mathcal{I}}^m]$ for each process $\mathcal{I} \in \mathbb{H}$ by solving a maximum likelihood estimation (MLE) problem. To do so, it is easy to show that the resulting log-likelihood function is

$$\log[L(\boldsymbol{\lambda}_{\mathcal{I},0}, \boldsymbol{B}|\epsilon, \boldsymbol{\omega}, \omega)]$$

$$= \sum_{\mathcal{I} \in \boldsymbol{\mathcal{I}}} \sum_{t_i \in \mathcal{H}_{\mathcal{I}}(T)} \log \lambda_{\mathcal{I}}(\tau_i) - \sum_{\mathcal{I} \in \boldsymbol{\mathcal{I}}} \int_0^T \lambda_{\mathcal{I}}(t) dt. \tag{10}$$

where $\boldsymbol{\lambda}_{\mathcal{I},0} := [\lambda_{\mathcal{I}_1,0}, \lambda_{\mathcal{I}_2,0}, \dots, \lambda_{\mathcal{I}_n,0}]$ and $\boldsymbol{B} \in \mathbb{R}^{n \times m}$ with $\boldsymbol{B}_{l,i} = \beta_{\mathcal{I}_l}^i$ are the variables to be estimated.

To incorporate the effect of competing processes, we further restrict $\lambda_{\mathcal{I}}(t)$ following Eq. (9), by first splitting the interval $[0, T)$, into L sets of small, equal, and disjoint subintervals $[0, \sigma), [\sigma, 2\sigma), \dots, [(L-1)\sigma, \sigma)$, where $\sigma = T/L$, and then imposing the following constraints, where $T_i = i\sigma$:

Whenever, $N_{\mathcal{I}}[T_i, T_{i+1}) \geq N_{\mathcal{I}'}[T_i, T_{i+1})$,

$$\int_{T_i}^{T_{i+1}} (\lambda_{\mathcal{I}}(t) - \lambda_{\mathcal{I}'}(t)) dt \geq 1; \ \mathcal{I}, \mathcal{I}' \in \boldsymbol{\mathcal{I}}, \ 0 \leq i \leq L - 1.$$

Similar to SVM [31], such a hard-margin approach often may lead to an infeasible solution. Therefore, we introduce slack variables

$$\zeta_{\mathcal{I},\mathcal{I}'}^i = \max(0, 1 - y_{\mathcal{I},\mathcal{I}'}^i \int_{T_i}^{T_{i+1}} [\lambda_{\mathcal{I}}(t) - \lambda_{\mathcal{I}'}(t)] dt) \tag{11}$$

with

$$y_{\mathcal{I},\mathcal{I}'}^i = \text{sign}(N_{\mathcal{I}}[T_i, T_{i+1}) - N_{\mathcal{I}'}[T_i, T_{i+1})) \tag{12}$$

and cast the problem as

$$\max_{\boldsymbol{\lambda}_{\mathbb{H},0}, \boldsymbol{B}} \log[L(\boldsymbol{\lambda}_{\mathbb{H},0}, \boldsymbol{B}|\epsilon, \boldsymbol{\omega}, \omega)] - C \sum_{i=0}^{L-1} \sum_{\mathcal{I},\mathcal{I}' \in \mathbb{H}} \zeta_{\mathcal{I},\mathcal{I}'}^i$$

$$y_{\mathcal{I},\mathcal{I}'}^i \int_{T_i}^{T_{i+1}} (\lambda_{\mathcal{I}}(t) - \lambda_{\mathcal{I}'}(t)) dt \geq 1 - \zeta_{\mathcal{I},\mathcal{I}'}^i \tag{13}$$

$$\forall \mathcal{I}, \mathcal{I}' \in \mathbb{H} \text{ and } 0 \leq i \leq L - 1.$$

Note that the above problem is convex and thus can be solved efficiently. We call this framework, *Large-Margin self-exciting Point Process* (LMPP), since it incorporates the variations in ranking by increasing the popularity-margins of competing hashtags while maximizing the corresponding log-likelihood.

2.2.1 Popularity Forecasting

Our goal here is to develop efficient methods that leverage our model to forecast a hashtag's popularity at a given time t. In the context of our model, we aim to compute $N^*_{\mathcal{I}}(t) = \mathbb{E}_{\mathcal{H}_{\mathcal{I}}(t)}[N_{\mathcal{I}}(t)]$, the expected value of total retweet counts of all tweets for a given hashtag $\mathcal{I}$ as:

$$N^*_{\mathcal{I}}(t) = \mathbb{E}_{\mathcal{H}_{\mathcal{I}}(t)}[N_{\mathcal{I}}(t)] = \mathbb{E}_{\mathcal{H}_{\mathcal{I}}(t)}\left[\int_0^t \tilde{\lambda}_{\mathcal{I}}(t)dt\right].$$

2.3 Competing Recurrent Point Process (CRPP)

Here, we formulate CRPP, the proposed model (see Fig. 1) for visibility dynamics in information diffusion. In this section, we describe them in detail, beginning with a brief review of temporal point process that characterizes the information diffusion dynamics considered in this paper.

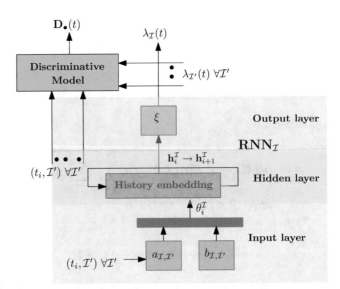

Fig. 1 The neural network architecture of CRPP for an process $\mathcal{I}$. At a high level, it shows the generator modeled as RNN and its interaction with discriminator. The current event (t_i) for an process $\mathcal{I}'$ is fed to the generator. It undergoes multiple transformations in the input layer and generates embedded signals $\theta_i^{\mathcal{I}}$, which are fed into the hidden layer where the states are computed recursively to learn the proper representation to capture the appropriate nonlinear influences of past events. The computed state $\mathcal{I}_i^{\mathcal{I}}$ is fed to the output layer which computes the conditional intensity $\lambda_{\mathcal{I}}(t)$ at time t. The output intensities along with input timestamps further go as inputs to the discriminative module which captures the ranking dynamics of the process

Temporal Point Process At the very outset, any online information diffusion process consists of asynchronous arrival of events (messages and reviews) which contain the posting times and the content of the event. Then we represent these event times as a temporal point process which is a stochastic process consisting of a sequence of discrete events localized in time. Formally, given a set of information diffusion processes $\mathcal{I}$, we define $\mathcal{H}_\mathcal{I}(t)$, also called as the history of $\mathcal{I}$ until time t, as the sequence of corresponding events for the process $\mathcal{I}$ until time t, i.e., $\mathcal{H}_\mathcal{I}(t) := \{t_j < t | \text{an event in } \mathcal{I} \text{ occurs at time } t_j\}$. It can also be realized by a counting process $N_\mathcal{I}(t) \in \{0\} \cup \mathbb{Z}^+$ that counts the number of events in $\mathcal{I}$ during $[0, t)$, i.e., $N_\mathcal{I}(t) = |\mathcal{H}_\mathcal{I}(t)|$. $\{0\} \cup \mathbb{Z}^+$ is used as the domain instead of $\mathbb{Z}$ since $N_\mathcal{I}(t)$ is a count which is either zero or a positive integer, whereas $\mathbb{Z}$ is the set of all integers (positive, negative, and zero). Then, we characterize the event rate for $\mathcal{I}$ using the conditional intensity function $\lambda_\mathcal{I}(t)$ that in turn is also associated with the conditional probability of observing an event in the infinitesimal time interval $[t, t + dt)$, given the history $\mathcal{H}(t)$ of use events from all the processes until time t:

$$\mathbb{P}(\text{An event in } \mathcal{I} \text{ occurs in } [t, t + dt) | \mathcal{H}_\mathcal{I}(t)) = \lambda_\mathcal{I}(t) \, dt$$

$$\text{i.e.,} \mathbb{E}[dN(t)|\mathcal{H}(t)] = \lambda_\mathcal{I}(t)dt. \tag{14}$$

Here, $dN_\mathcal{I}(t)$ gives the number of product use events in the infinitesimal time-window $[t, t + dt)$ and $\lambda_\mathcal{I}(t)$ gives the instantaneous event intensity. Therefore, the generative modeling of the event diffusion process, i.e., modeling the generation of t_i's, essentially boils down to an appropriate formulation of $\lambda_\mathcal{I}(t)$. However, most of the existing approaches consider $\lambda_\mathcal{I}(t)$ having a fixed parametric form with linear dynamics, e.g., Poisson process, Hawkes process [1, 10, 17, 34], etc. In contrast to these works, in this paper, we seek to learn the functional form without any restrictive imposition that in turn provides a predictive prowess to our model.

2.3.1 CRPP: Competing Recurrent Point Process

The visibility dynamics of a process has two major components: (1) the dynamics of collective flow of product/hashtag use events, and (2) the variation of popularity ranking during competition, which are detailed hereinafter, and also sketched in Fig. 1.

CRPP$_{\text{Gen}}$, the Generative Module for CRPP Given a set of processes $\mathcal{I}$, and the histories $\mathcal{H}_\mathcal{I}(t), \mathcal{I} \in \mathcal{I}$, the generic form of joint generative dynamics is given by

$$\lambda_\mathcal{I}(t) = f_\mathcal{I}(\cup_{\mathcal{I} \in \mathcal{I}} \mathcal{H}_\mathcal{I}(t)), \tag{15}$$

where f can be any arbitrary nonlinear function, as opposed to the restrictive forms. In this work, we aim to model this function using recurrent neural network (RNN). RNNs are a family of feedforward neural architectures with some auxiliary edges, called recurrent edges, that connect the current signals from hidden state to the network as the future inputs at the very next time-step. This recursive structure creates an inbuilt memory, and therefore captures the complex influence of memory from the previous events, which in turn helps to accurately model a temporal point process [6].

At a very high level, our model assigns one RNN (RNN$_{\mathcal{I}}$) per each product $\mathcal{I}$, which takes previous posting times $t_i \in \cup_{\mathcal{I}\in\mathcal{I}}\mathcal{H}_{\mathcal{I}}(t)$ as inputs, and outputs the intensity $\lambda_{\mathcal{I}}(t)$ for the next event as shown in Fig. 1. In this process, the hidden states of RNN$_{\mathcal{I}}$ capture the history $\mathcal{H}_{\mathcal{I}}(t)$ using the embeddings $h_k^{\mathcal{I}}$ which are recursively computed from the previous embedding vectors $h_{k-1}^{\mathcal{I}}$ as well as the signals derived from current input timings. In the following, we elaborate more on our architecture which has three layers: (1) input layer, (2) hidden layer, and (3) output layer.

Input Layer For each process $\mathcal{I}$, the *input layer* takes the use event arrival times $t_i \in \mathcal{H}_{\mathcal{I}}$ as inputs and turns them into suitable signals, thereby preparing the stage for functioning of the hidden layer. That said, upon arrival of an event of $\mathcal{I}'$ at time t_i, the input layer converts it into a signal $\theta_i^{\mathcal{I}} = a_{\mathcal{I},\mathcal{I}'}t_i + b_{\mathcal{I},\mathcal{I}'}$. Here $a_{\mathcal{I},\mathcal{I}'}$ and $b_{\mathcal{I},\mathcal{I}'}$ offer a direct measure of influence of $\mathcal{I}'$ on $\mathcal{I}$. In a marked departure from the existing works [30] assuming independent dynamics for concurrent processes, our model enforces a dependence structure in event flow.

Hidden Layer The computed vectors $\theta_i^{\mathcal{I}}$, as well as the timings t_i, are fed into the recurrent unit where the hidden state computes $h_i^{\mathcal{I}}$ using the previous embedding $h_{i-1}^{\mathcal{I}}$ in the following way:

$$h_i^{\mathcal{I}} = g_w \Big(l_{\mathcal{I}} h_{i-1}^{\mathcal{I}} + \underbrace{\beta_{\mathcal{I}} e^{-\frac{\omega t_i}{k_i}}}_{\substack{\text{influence of event} \\ \text{magnitude}}} + \gamma_{\mathcal{I}}^T \theta_i^{\mathcal{I}} \Big). \tag{16}$$

Here g_w is a nonlinear function realized by the neural network with parameters w. The second term in Eq. (16) contains k_i which is the popularity index of event occurring at time t_i. This term acts as a reinforcement factor for influence from previous events. For example, if k_i is high (low) the effect of $e^{-\omega t_i/k_i}$ becomes high (low). In other words, the popular process would influence another process more than a not-so-popular process. In the case of the Twitter, k_i can be looked upon as a normalized measure of $(N_i / \sum_j N_j)$ the number of retweets of the tweet i in the whole interval $[0, T)$. Such a term encourages "hashtag-tweet reinforcement" (using the term coined by Samanta et al. [26]), where a popular tweet drives the flow of the underlying hashtag, rather than a not-so-popular tweet. In case of the scenario of the products at Amazon, where a user review corresponds to a product use event, k_i would denote the rating of the review, which is on a scale of 1–5 stars. In this case, the event magnitude models what may be analogously called "product-review reinforcement," where a review with a higher star rating boosts the popularity of the product, leading to higher sales and correspondingly more product use events. To this end, an event, either a popular or a non-popular tweet or a review with a high or low rating, with a high or small value of k_i, boosts or trims the values of $h_i^{\mathcal{I}}$ that later on increases or reduces the message intensity $\lambda_{\mathcal{I}}(t)$ which is generated in the output layer.

Output Layer It computes the intensity for each process $\mathcal{I}$ as:

$$\lambda_{\mathcal{I}}(t) = \exp(\boldsymbol{\xi}(\boldsymbol{\omega}^T \boldsymbol{h}_i^{\mathcal{I}} + b(t - t_i) + c)), \tag{17}$$

where t_i is the time of the last observed event prior to t, b is a weight parameter, and c is a bias parameter. Such an exponential ensures a positive value of $\lambda_{\mathcal{I}}(t)$ which may be violated otherwise during training process. Here, $\boldsymbol{\xi}$ is also realized using a neural network. It may be noted that this form of the conditional intensity is heavily inspired by Du et al. [6], which also takes account of the influence of the last observed event.

CRPP$_{\text{Discrim}}$, the Discriminative Module of CRPP Competition between two processes manifests relative variation in their popularity rankings over time. To model this ranking dynamics, we first split the interval $[0, T)$, into L sets of small, equal, and disjoint subintervals $[0, T_s), [T_s, 2T_s), \ldots, [(L-1)T_s, T)$, where $T_s = T/L$. Then, we define

$$y_{\mathcal{I},\mathcal{I}'}^i = \mathbb{I}(N_{\mathcal{I}}[iT_s, (i+1)T_s) > N_{\mathcal{I}'}[iT_s, (i+1)T_s)),$$

$$\Delta_{\mathcal{I},\mathcal{I}'}^i = \int_{iT_s}^{(i+1)T_s} (\lambda_{\mathcal{I}}(t) - \lambda_{\mathcal{I}'}(t))dt. \tag{18}$$

Here $\mathbb{I}[x]$ is the indicator function which is equal to 1 (0) if x is true (false). Furthermore, $y_{\mathcal{I},\mathcal{I}'}^i = 1$ (0) indicates that the *observed* popularity of $\mathcal{I}$ is greater (less) than that of $\mathcal{I}'$ during $[iT_s, (i+1)T_s)$. On the other hand, $\Delta_{\mathcal{I},\mathcal{I}'}^i$ *estimates* the difference in popularity between $\mathcal{I}$ and $\mathcal{I}'$. Now, for each interval $[iT_s, (i+1)T_s)$, we employ several ranking loss functions to capture the ranking variations of processes (i.e., variations of $y_{\mathcal{I},\mathcal{I}'}^i$ across $0 \leq i < L$). These loss functions are often used in learning to rank in the context of information retrieval [12].

Unbiased Ranking Model A simple unbiased ranking function may be derived following a large margin approach. Suppose we are given two processes $\mathcal{I}_1, \mathcal{I}_2 \in \mathcal{I}$ and time instances t_s and t_f $(t_s < t_f)$

$$N_{\mathcal{I}_1}[t_s, t_f) \geq N_{\mathcal{I}_2}[t_s, t_f) \implies \int_{t_s}^{t_f} \lambda_{\mathcal{I}_1}(t)dt \geq \int_{t_s}^{t_f} \lambda_{\mathcal{I}_2}(t)dt.$$

Here, $N_{\mathcal{I}}[t_s, t_f)$ indicates the number of use events of process $\mathcal{I}$ occurring in the interval $[t_s, t_f)$. In words, if the popularity (number of use events) of $\mathcal{I}_1$ is more than $\mathcal{I}_2$, so are their estimates $\int_{t_s}^{t_f} \lambda_*(t)dt$. There are several ways to capture such scenarios. The easiest way is to plug them as constraints while estimating the parameters using the likelihood function. Another plausible direction is to follow the probabilistic approach that we adopt in our proposed model. We wish to minimize the number of instances, where $y^i_{\mathcal{I},\mathcal{I}'}$ and $\boldsymbol{\Delta}^i_{\mathcal{I},\mathcal{I}'}$ are not in agreement. That is, we maximize $\prod_{\mathcal{I},\mathcal{I}'\in\mathcal{I}}\left(1 - y^i_{\mathcal{I},\mathcal{I}'}\mathbb{I}[\boldsymbol{\Delta}^i_{\mathcal{I},\mathcal{I}'} < 0] - (1 - y^i_{\mathcal{I},\mathcal{I}'})\mathbb{I}[\boldsymbol{\Delta}^i_{\mathcal{I},\mathcal{I}'} > 0)\right)$ which, however, is not differentiable. By approximating it using sigmoids we define the discriminative ranking loss function as:

$$D_{\text{unbiased}}(i) := \prod_{\mathcal{I},\mathcal{I}'\in\mathcal{I}}\left(1 - y^i_{\mathcal{I},\mathcal{I}'}\sigma[W; -\boldsymbol{\Delta}^i_{\mathcal{I},\mathcal{I}'}] - (1 - y^i_{\mathcal{I},\mathcal{I}'})\sigma[W; \boldsymbol{\Delta}^i_{\mathcal{I},\mathcal{I}'}]\right).$$
(19)

An unbiased ranking model is useful when the variation of visibility across $\mathcal{I}$ is uniform, i.e., when for a randomly chosen process $\mathcal{I}$, the numbers of more visible and less visible processes remain nearly same. For a skewed visibility dynamics where there are only few visible products, we employ more complex ranking losses, e.g., precision, recall, area under ROC curve (AUC), etc., that are more reliable measures in ranking imbalanced data.

Precision Precision measures how accurately the estimated rank of a process indicates its actual visibility. Given a process $\mathcal{I}$, Precision($\mathcal{I}, i$) is defined as the probability that it is actually more visible than a randomly selected process $\mathcal{I}'$, given that $\mathcal{I}$ is estimated to be more visible than $\mathcal{I}'$, during the interval $[iT_s, (i+1)T_s)$. Formally, it can be written as:

$$\text{Precision}(\mathcal{I}, i) := \frac{\sum_{\mathcal{I}'} y^i_{\mathcal{I},\mathcal{I}'}\mathbb{I}[\boldsymbol{\Delta}^i_{\mathcal{I},\mathcal{I}'} > 0]}{\sum_{\mathcal{I}'}\mathbb{I}[\boldsymbol{\Delta}^i_{\mathcal{I},\mathcal{I}'} > 0]}$$
(20)

which again is a non-smooth ranking function that can be approximated using sigmoid, and the corresponding discriminative loss $D_{\text{Precision}}(\mathcal{I}, i)$ becomes

$$D_{\text{Precision}}(\mathcal{I}, i) = 1 - \frac{\sum_{\mathcal{I}'} y^i_{\mathcal{I},\mathcal{I}'}\sigma(W; \boldsymbol{\Delta}^i_{\mathcal{I},\mathcal{I}'})}{\sum_{\mathcal{I}'}\sigma(W; \boldsymbol{\Delta}^i_{\mathcal{I},\mathcal{I}'})}.$$
(21)

Recall Given a process $\mathcal{I}$, Recall($\mathcal{I}$) is defined as the probability that it is estimated to have a higher rank than $\mathcal{I}'$, given that it actually ranks higher than $\mathcal{I}'$, during the time interval $[iT_s, (i+1)T_s)$. Formally, it can be written as:

$$\text{Recall}(\mathcal{I}, i) := \mathbb{P}(\boldsymbol{\Delta}^i_{\mathcal{I},\mathcal{I}'} > 0 | y^i_{\mathcal{I},\mathcal{I}'} > 0),$$
(22)

which is turned into a differentiable discriminative loss

$$D_{\text{Recall}}(\mathcal{I}, i) := 1 - \frac{\sum_{\mathcal{I}'} y^i_{\mathcal{I},\mathcal{I}'}\sigma(W; \boldsymbol{\Delta}^i_{\mathcal{I},\mathcal{I}'})}{\sum_{\mathcal{I}'} y^i_{\mathcal{I},\mathcal{I}'}}.$$
(23)

AUC Before going to AUC, we first connect the idea of ROC (receiver operating characteristics) in the context of product (hashtag) visibility. Suppose, we obtain a ranked list of $\mathcal{I} \in \boldsymbol{\mathcal{I}}$ according to the value of $\int_{iT_s}^{(i+1)T_s} \lambda_{\mathcal{I}}(t)dt$. Then at each position $k \in \{1, \ldots, |\boldsymbol{\mathcal{I}}|\}$, we compute the true positive rate $\text{TPR}_k := \mathbb{P}_k(\boldsymbol{\Delta}_{\mathcal{I},\mathcal{I}'}^i > 0, y_{\mathcal{I},\mathcal{I}'}^i > 0)$ for two randomly selected products (hashtags) $\mathcal{I}$ and $\mathcal{I}'$ within rank k, as well as the corresponding false positive rate $\text{FPR}_k := \mathbb{P}_k(\boldsymbol{\Delta}_{\mathcal{I},\mathcal{I}'}^i < 0, y_{\mathcal{I},\mathcal{I}'}^i > 0)$. The variation of TPR_k (Y axis) with FPR_k (X axis) provides the ROC curve in our context. The area under the ROC curve offers an important measure in the context of any ranking paradigm. In fact, it is well known [12] that AUC during the interval $[iT_s, (i+1)T_s)$ is the same as

$$\text{AUC}(i) = 1 - \frac{\sum_{\mathcal{I},\mathcal{I}'} y_{\mathcal{I},\mathcal{I}'}^i \mathbb{I}[\boldsymbol{\Delta}_{\mathcal{I},\mathcal{I}'}^i < 0]}{|\boldsymbol{\mathcal{I}}|^2}. \tag{24}$$

Hence the underlying discriminative loss becomes

$$\boldsymbol{D}_{\text{AUC}}(i) = \frac{\sum_{\mathcal{I},\mathcal{I}'} y_{\mathcal{I},\mathcal{I}'}^i \boldsymbol{\sigma}(\boldsymbol{W}; -\boldsymbol{\Delta}_{\mathcal{I},\mathcal{I}'}^i)}{|\boldsymbol{\mathcal{I}}|^2}. \tag{25}$$

When the nature of the ground truth is unknown, then we recommend using AUC as the discriminative loss a priori since it is considered to be a more reliable measure than precision or recall [12].

2.3.2 Inference

Consider a set of processes $\boldsymbol{\mathcal{I}} = \{\mathcal{I}\}$, associated with a collection of use events $\mathcal{H}_{\mathcal{I}}(T) = \{t_i\}$ for each process $\mathcal{I}$ during a time period $[0, T)$. Using these timestamps, we attempt to infer the underlying generative process, as well as the parameters of the discriminative ranking model. To learn the parameters, we maximize the log-likelihood for the recorded events $\cup_{\mathcal{I} \in \boldsymbol{\mathcal{I}}} \mathcal{H}_{\mathcal{I}}(T)$ combined with the ranking losses. Here, the log-likelihood function is

$$\log[L(a_{\bullet,\bullet}, b_{\bullet,\bullet}, w, \xi, l_{\bullet}, \beta_{\bullet}, \gamma_{\bullet}, \omega_{\bullet}, b, c)]$$

$$= \sum_{\mathcal{I} \in \boldsymbol{\mathcal{I}}} \sum_{t_i \in \mathcal{H}_{\mathcal{I}}(T)} \log \lambda_{\mathcal{I}}(t_i) - \sum_{\mathcal{I} \in \boldsymbol{\mathcal{I}}} \int_0^T \lambda_{\mathcal{I}}(t)dt. \tag{26}$$

All the variables in the arguments of $L(.)$ are the parameters of the neural networks in the generative model. To incorporate the effect of competing products, we again split the interval $[0, T)$, into L sets of small, equal, and disjoint subintervals $[0, T_s), [T_s, 2T_s), \ldots, [(L-1)T_s, T)$, where $T_s = T/L$, where for each interval, we have a corresponding discriminative loss $\boldsymbol{D}_{\text{Measure}}(\mathcal{I}, i)$, for Measure $\in$ {Precision, Recall}, or $\boldsymbol{D}_{\text{Measure}}(i)$ where Measure $\in$ {Unbiased, AUC}. Finally, we compute

the parameters by minimizing the generative loss (negative log-likelihood) simultaneously with the discriminative loss in the following way:

$$\min_{W,G} -\log L(\cup_{\mathcal{I}\in\mathcal{I}}|G) + \sum_{i=1}^{L}\sum_{\mathcal{I}\in\mathcal{I}}\log(D_{\text{Measure}}(\mathcal{I}, i)).$$

Here G is given by the neural parameters of the generative model. In the above
optimization problem, the discriminative model D reinforces and corrects the
generative model G, until the parameters are learned in such a way that the intensity
functions respect the observed ranking dynamics across time.

3   Experiments

In this section, we provide a comprehensive evaluation of LMPP and CRPP
across a diverse type of real-world datasets gathered from Twitter and Amazon.
More specifically, we test the utility of our proposed models by investigating
how accurately (1) they can model the underlying competition mechanism and (2)
they can forecast the dynamics of the individual information diffusion process. To
elaborate them further, in the following, we first give a brief descriptions of the
datasets used, the evaluation protocol, the baseline methods, and then provide a
detailed comparative analysis of LMPP and CRPP against several baselines.

3.1   Datasets

We evaluate our proposed models on four datasets collected from Twitter and
six datasets collected from Amazon, as described below and also summarized in
Table 1. All these datasets are publicly available.

Twitter Datasets Out of the four datasets used for our evaluation, one dataset
Halloween was gleaned from a parent dataset consisting of tweets posted in
November 2012, which was used in [20, 21] and the rest *Nepal-Earthquake*, *Dem-
Primary*, and *BBD* were used in [28]. All of these datasets comprise messages
collected from a diverse type of events, e.g., sports, entertainment, movies, etc.,
during 2–3 weeks around the corresponding event. The hashtags selected from these
messages respect two criteria: (1) They have a significant number of concurrent and
related hashtags with a large number of tweets. (2) They show substantial deviations
in popularity ranking across the timeline. In this context, we measure rank diversity
of a hashtag, defined as the fraction of times its rank has changed. If out of a total
of I time-windows, a hashtag $H \in \mathbb{H}$ has changed its rank k times, then rank
diversity$(H) = k/I$. The mean rank diversity of a dataset provides a measure of
aggregated rank fluctuations of the underlying hashtags (see Table 1). Within the
same dataset, different diffusion processes correspond to the reviews of different
hashtags that are competing with each other.

Table 1 Statistics of datasets collected

Datasets	Duration, T	Example-hashtags/brands	$\sum_{\mathcal{I}} |\mathcal{H}_{\mathcal{I}}(T)|$	$\mathbb{E}[|\mathcal{H}_{\mathcal{I}}(T)|]$	$\sigma(k_i)$
Halloween	Nov 1, 2012 to Nov 30, 2012	#Halloween2012	38,719	22.38	0.43
Nepal-earthquake	Apr 25 to May 1, 2015	#kathmanduquake,#nepalrelief	32,613	1304.52	0.70
Dem-Primary	Feb to June, 2016	#makeamericagreatagain,#PresumptiveNominee	21,746	1449.73	0.72
BBD	Oct 6 to Oct 8, 2014	#aslidealsonebay,#bigbillionsale	67,399	3369.95	0.51
Amazon-SP	Aug, 2001 to Oct, 2014	Huawei Mate, Honor 7X	2174	217.4	0.48
Amazon-LAP	Aug, 2001 to Oct, 2014	Asus Vivobook, Acer Flagship	2112	211.2	0.47
Amazon-REF	Aug, 2001 to Oct, 2014	Midea, RCA, Danby	2872	287.2	0.49
Amazon-WM	Aug, 2001 to Oct, 2014	Kenmore, Haier, Giantex	2684	268.4	0.45
Amazon-AC	Aug, 2001 to Oct, 2014	Honeywell, Frigidaire, LG	2625	262.5	0.48
Amazon-TV	Aug, 2001 to Oct, 2014	TCL, Sceptre, LG	1883	188.3	0.46

The first four datasets are collected from Twitter and the rest are gathered from Amazon. $\sigma(k_i)$ denotes mean rank diversity

Amazon We used the raw product-review data of Amazon used in [9, 19]. In particular, we gathered the unix review timestamps of products of well-known brands, with at least 100 reviews each, belonging to several highly specific product categories of high value and long life-expectancy, e.g., laptops, refrigerators, washing machines, etc., which we modeled as a temporal process. It leads to six different datasets corresponding to six product categories: Smartphones (*Amazon-SP*), Laptops (*Amazon-LAP*), Refrigerators (*Amazon-REF*), Washing Machines (*Amazon-WM*), Air Conditioners (*Amazon-AC*), and Televisions (*Amazon-TV*). Within the same product category, different diffusion processes correspond to the reviews of different products that are competing with each other, probably due to brand contest between products.

3.2 Evaluation Protocol

Training and Testing At the very outset, given a set of I processes, i.e., I sets of event-streams, we split the entire set of events into training and test set, where training comprises of first 80% of events. More specifically, suppose the total window of observation is $(0, T]$. Then, we divide it into training and test sets as $(0, T_{\text{train}}]$ and $(T_{\text{train}}, T]$, respectively, so that $|\mathcal{H}_{\mathcal{I}}(T_{\text{train}})| = 0.8|\mathcal{H}_{\mathcal{I}}(T)|$. In addition, to probe the variation in ranking, we divide the entire training window $(0, T_{\text{train}}]$ in a total of 100 equally spanned time intervals to obtain the popularity ranked lists as ground truths. That said, in the line of Eq. (18), we define $T_s = T/100$ so that the ith such interval would be $[iT_s, (i + 1)T_s)$. The popularity of a diffusion process $\mathcal{I}$ (i.e., the popularity of a hashtag) in this interval is given by INTERVALPOP($\mathcal{I}, i$) = $N_{\mathcal{I}}[iT_s, (i + 1)T_s)$. When we say that an item $\mathcal{I}$ is more popular than an item $\mathcal{I}'$ in $[iT_s, (i + 1)T_s)$, we mean that INTERVALPOP($\mathcal{I}, i$) > INTERVALPOP($\mathcal{I}', i$). In case INTERVALPOP($\mathcal{I}, i$) = INTERVALPOP($\mathcal{I}', i$), their ranks would be equal. However, when duplicate ranks are not allowed, ties can be broken arbitrarily. Thus, the processes $\mathcal{I} \in \mathcal{I}$ can be ranked according to interval popularity in this manner to produce a ranked list $R_{\mathcal{I}}([iT_s, (i + 1)T_s))$ for each interval $[iT_s, (i + 1)T_s)$. Using the observations in the training window $(0, T_{\text{train}}]$, as well as the ranked list $R_{\mathcal{I}}([iT_s, (i + 1)T_s))$ in each interval $[iT_s, (i + 1)T_s)$, we train our model CRPP and estimate all the parameters $(a_{\bullet,\bullet}, b_{\bullet,\bullet}, w, \xi, l_{\bullet}, \beta_{\bullet}, \gamma_{\bullet}, \omega_{\bullet}, b, c)$. Finally, we use these parameters to predict the events in the test set. In this context, we used Ogata's thinning algorithm [25] to sample from the intensity function.

   We evaluate our proposed methods from the following two perspectives—that in turn test the two key modeling machineries of LMPP and CRPP.

Forecasting Performance Here, we aim to measure how effectively LMPP and CRPP predict the dynamics of the diffusion process in future. To do that, we first predict future popularity of processes by computing the estimated number of events. Apart from that, we also try to predict the timestamps of the next event.

Rank Prediction of the Competing Processes To measure how accurately our models are able to capture the competition between the diffusion processes, we try to predict the ranks of the individual processes in the subintervals of the test set.

Evaluation Metrics We measure the efficacy of our models using the following metrics. The first set of metrics, i.e., $MAPE_N$ and $MAPE_t$, measures the forecasting ability of the algorithms, while the rest, i.e., SRCC, and precision and recall, indicate how accurately our model captures the competition between the diffusion processes.

MAPE $_N$ We report mean absolute percentage error ($MAPE_N$) to measure the popularity prediction errors defined as below (where $\hat{N}_\mathcal{I}$ is the predicted count and $N_\mathcal{I}$ is the actual count)

$$\text{MAPE}_N(\mathcal{I}) = \frac{1}{M_\mathcal{I}} \sum_{t_i \in \mathcal{H}_\mathcal{I}(T_{\text{train}}, T]} \left| \frac{\hat{N}_\mathcal{I}(t_i) - N_\mathcal{I}(t_i)}{N_\mathcal{I}(t_i)} \right|. \tag{27}$$

MAPE $_t$ $MAPE_t$ reports the mean deviation of the estimated timestamps of the next future events, from their actual times. Formally, if $\hat{t}_i$ is the predicted time and t_i is the actual time

$$\text{MAPE}_t(\mathcal{I}) = \frac{1}{M_\mathcal{I}} \sum_{t_i \in \mathcal{H}_\mathcal{I}(T_{\text{train}}, T]} \left| \frac{\hat{t}_i - t_i}{t_i} \right|. \tag{28}$$

In the above two metrics, $M_\mathcal{I}$ is the number of events of $\mathcal{I}$ in test set. Finally for each dataset, we report $MAPE_N$ and $MAPE_t$ which are calculated as averages of all $MAPE_N(\mathcal{I})$ and $MAPE_t(\mathcal{I})$ over all possible $\mathcal{I} \in \mathbf{\mathcal{I}}$.

SRCC To measure how accurately we capture the ranks of the processes, we measure Spearman's rank correlation coefficient (SRCC) for predicted ranklist $\hat{R}_\mathcal{I}$ with the ground truth list $R_\mathcal{I}$ as below:

$$\rho(\hat{R}_\mathcal{I}, R_\mathcal{I}) = \frac{\mathbf{Cov}(\hat{R}_\mathcal{I}, R_\mathcal{I})}{\sqrt{\mathbf{Var}(\hat{R}_\mathcal{I})\mathbf{Var}(R_\mathcal{I})}}. \tag{29}$$

Average Precision and Recall We also measure the ability of the models to predict the instances where a particular diffusion process suddenly gains popularity (is significantly promoted in rank) or suddenly loses popularity (is significantly demoted in rank). Formally, in two consecutive intervals, if the change in the rank of an item is more than half of total no. of items, we call it a *jump*. That is, given the rank of a hashtag $\mathcal{I} \in \mathbf{\mathcal{I}}$ be $\rho_{\mathcal{I},[t_i,t_{i+1})}$ at time interval $[t_i, t_{i+1})$, if $|\rho_{\mathcal{I},[t_i,t_{i+1})} - \rho_{\mathcal{I},[t_{i-1},t_i)}| \geq |\mathbf{\mathcal{I}}|/2$, it is considered a jump. This jump value is an average value, between the maximum jump, i.e., $|\mathbf{\mathcal{I}}|$, and minimum jump, i.e., 0. It was chosen to provide an average-case analysis. Recall measures the proportion of real jumps which are correctly identified by an algorithm, while precision measures the fraction of cases where the jump predicted by an algorithm is actually observed in real data.

3.3   Baselines

Our second proposal, CRPP, which contains a discriminative model apart from a generator, operates over a variety of ranking functions, i.e., unbiased ranking loss, precision, recall, and AUC. We call the corresponding derivatives of our model $\text{CRPP}_{\text{Unbiased}}$, $\text{CRPP}_{\text{Precision}}$, $\text{CRPP}_{\text{Recall}}$, and CRPP_{AUC} respectively. We compare LMPP and CRPP (all its variants) with several state-of-the-art baselines, viz.[1]

(i)   **Wasserstein GAN Temporal Point Process (WGANTPP)** [32]: It is a deep generative model for temporal point process, where the deep neural network is constructed using a Wasserstein generative adversarial setting.

(ii) **Reinforcement Poisson Process (RPP)**[10]: RPP computes the popularity of a diffusion process as the following:

$$c^d(t) = (m + n_d)e^{\lambda_d^*(F_d(t;\theta_d*) - F_d(T;\theta_d^*))} - m,$$

where $m = |\mathcal{I}|$, $n_d = \sum_{\mathcal{I}} |\mathcal{H}_{\mathcal{I}}(T)|$ is the entire number of events, $\lambda_d(t)$ indicates an intrinsic attractiveness of the processes, $F_d(t;\theta_d) = \int_0^t f_d(t;\theta_d)dt$ with $f_d(t;\theta_d)$ as a relaxation function, and $i_d(t)$ is a reinforcement factor. All these parameters are learned using the training data.

(iii) **Hawkes Process**[1]: Hawkes process which is considered as the workhorse of any temporal point process based diffusion models is given by

$$\lambda(t) = \mu + \alpha \int_0^t \exp(-\beta t)dN(t).$$

Here β is the decay factor and μ and α are trainable parameters. A simple Hawkes process offers analytical solution for the predicted time $\hat{t}_i$ as well as the predicted count $\hat{N}$, thereby giving accurate forecasting formulas for future events.

(iv) **SEISMIC** [34]: SEISMIC only estimates the final size of an information cascade in the context of networked diffusion process. It is derived using Hawkes process, which characterizes the events by the re-share probability. The prediction function is defined as

$$\hat{R}_\infty(t) = R_t + \alpha_t \frac{\hat{p}_t(N_t - N_t^e)}{1 - \gamma_t \hat{p}_t n_*}, 0 < \alpha_t, \gamma_t < 1, \tag{30}$$

where p_t is the infectiousness, N_t^e—effective cumulative degree of re-sharers by time t, and N_t—cumulative degree of re-sharers by time t.

(v)   **SpikeM** [18]: SpikeM characterizes the empirical observations of temporal behavior of popularity in social media. It models the user behavior as:

---

[1] Viz. stands for videlicet, a Latin word meaning "namely."

$$\Delta B(n+1) = p(n+1)(U(n)\sum_{t=n_b}^{n}(\Delta B(t) + S(t)).f(n+1-t) + \hat{\epsilon})$$

$B(n)$ and $U(n)$ give the number of infected (who have tweeted on an event) and uninfected (who have not tweeted on an event) users at time t_n, respectively. At any time, SpikeM aims to predict the number of infected users who have tweeted on an event.

Finally, to evaluate the performance of CRPP in the absence of the discriminator, we also consider as a baseline (vi) CRPP_{Gen}, which is our model of CRPP trained only to maximize the log-likelihood without any discriminator feedback. Since (vii) **RMTPP** (recurrent marked temporal point process) [6] is closely related to CRPP_{Gen}, we consider it separately from the other existing models so that we may use it as a baseline specifically to evaluate the performance of CRPP_{Gen}.

Further, in order to make a more fair comparison with CRPP, we tuned all baselines (except RMTPP) by applying an appropriate discriminative model to each while measuring performance on each metric, where we additionally minimize a ranking loss in addition to the existing MLE-based systems. In particular, while measuring MAPE_N, SRCC, and MAPE_t, the baselines were augmented with D_{unbiased} (Eq. (19)), $D_{\text{Precision}}$ (Eq. (21)), and D_{Recall} (Eq. (23)). While doing so, we made sure that the unique characteristics of the baselines were minimally disturbed. However, it must be noted that these models are not nearly as amenable to the discriminator as CRPP is naturally, and therefore are unable to take full advantage of it while predicting ranking dynamics.

3.4   Performance Comparison

We first compare the performance between the two proposed models, CRPP and LMPP and the baselines, taking the best-performing variant of CRPP as its representative. Thereafter we present a comparison among the individual variants of CRPP.

Performance in Popularity Forecasting Table 2 dissects a comparative analysis of different methods in terms of MAPE_N across all the datasets. It shows that CRPP performs best against all its competitors (including LMPP), by achieving the lowest MAPE_N. We observe that the performances of RPP, Hawkes, SEISMIC, and spikeM are substantially poor. This is because, these methods employ fixed parametric functional forms to model the dynamics of product use event propagation, which barely capture the complexity of the process. Moreover, they ideally operate only over individual events, not even on individual processes or a set of events. Consequently, they cannot capture the dynamics of information flow, as well as the competition between the concurrent processes. So, their predictive power turns out to be very poor, and more importantly, they perform inconsistently across the datasets. The second position among the models is shared by LMPP and WGANTPP. Despite

Table 2 MAPE$_N$ (%) of the proposed and baseline algorithms on all datasets with 20% held-out test set

Datasets	MAPE$_N$ (%)						
	CRPP	LMPP	WGANTPP	RPP	Hawkes	SEISMIC	SpikeM
Halloween	**12.45** (18.09%)	*15.20*	15.90	21.21	16.43	21.39	24.41
Nepal-earthquake	**6.70** (2.6%)	7.50	*6.88*	22.43	15.85	13.73	17.95
Dem-primary	**7.83** (6.00%)	*8.33*	8.35	11.52	11.96	26.09	19.12
BBD	**14.62** (5.06%)	*15.40*	15.45	19.45	16.23	18.03	20.89
Amazon-SP	**4.70** (17.69%)	6.10	*5.71*	13.23	18.35	12.83	22.15
Amazon-LAP	**4.21** (19.81%)	*5.25*	5.34	8.72	14.51	18.09	26.35
Amazon-REF	**4.85** (7.09%)	*5.22*	5.45	8.32	14.34	19.09	18.21
Amazon-WM	**4.79** (11.46%)	5.89	*5.41*	7.96	15.34	20.22	23.44
Amazon-AC	**4.82** (7.48%)	*5.21*	5.85	8.56	14.33	18.23	21.85
Amazon-TV	**4.53** (4.83%)	*4.76*	5.12	8.35	9.31	8.52	21.13

Bold (italics) indicate the best (second best) predictor. Here CRPP reports the performance of the best among the variants of our proposal. Numbers in bracket in the CRPP cells indicate the percentage improvement from the second best predictor

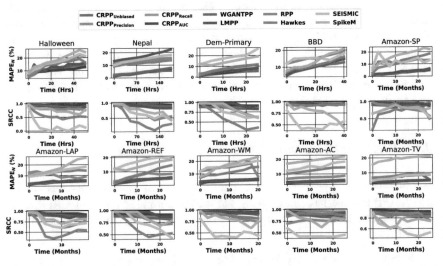

Fig. 2 Variation of popularity forecasting and rank prediction performance with time across all datasets for all variants of CRPP along with LMPP and the baselines

being a linear model, LMPP still models the product competition. On the contrary, WGANTPP can uncover the complexity of the nonlinear diffusion process, but it does not capture the ranking fluctuations well, despite being augmented with a ranking loss. CRPP, on the other hand suitably combines the diffusion dynamics as well as the competition in a unified way. As a result it performs consistently better than others.

We also plot the variation in forecasting performance (with respect to MAPE$_N$) with time in the test window for our models as well as the baselines in Fig. 2.

Table 3 MAPE$_t$ (%) of the best-performing variant of CRPP along with LMPP and the baselines on all datasets with 20% held-out test set

Datasets	MAPE$_t$				
	CRPP	LMPP	WGANTPP	RPP	Hawkes
Halloween2012	**5.43** (6.70%)	6.45	*5.82*	7.15	6.35
Nepal-earthquake	**6.41** (6.29%)	7.61	*6.84*	21.45	15.62
Dem-primary	**7.41** (12.31%)	*8.45*	8.56	11.32	11.77
BBD	**14.34** (5.72%)	*15.21*	15.36	19.35	16.65
Amazon-SP	**4.70** (14.39%)	6.16	*5.49*	6.21	6.35
Amazon-LAP	**4.31** (19.29%)	5.73	*5.34*	5.87	5.91
Amazon-REF	**4.83** (17.44%)	6.86	*5.85*	7.13	7.89
Amazon-WM	**4.79** (11.46%)	5.82	*5.41*	5.91	5.62
Amazon-AC	**4.62** (19.65%)	5.83	*5.75*	6.67	5.91
Amazon-TV	**4.53** (18.52%)	6.18	*5.56*	6.63	6.41

Bold (italics) indicate the best (second best) predictor. Numbers in bracket indicate the percentage improvement from the second best predictor

We observe that the prediction performance of all our models exhibits a graceful degradation as we move further into the test time (away from the last observed training instance). Whereas all the models show a general decline in performance with time, CRPP, LMPP, and WGANTPP are more robust to this effect.

Performance in Next-Event Prediction Table 3 reports MAPE$_t$-s for a comparison among our first model, LMPP, the best-performing variant of our second model, CRPP, and the baselines. We observe that among the baselines WGANTPP performs the best. We also observe that the general pattern of relative performance across the datasets for various models is similar to that observed in case of the metric MAPE$_N$. This may be quite natural since a higher MAPE$_N$ indicates a stronger event forecasting ability, which translates into accurate predictions for the event times.

Performance in Rank Prediction Table 4 depicts a comparative sketch of the predictive powers of different models in terms of SRCC. It reflects the ability of the algorithms to probe variations in popularity rankings during the competition process. We observe that CRPP performs substantially better than LMPP and all the baselines across all the datasets. Similar to MAPE$_N$, in this case too, both WGANTPP and LMPP perform reasonably well and second CRPP. We observe that, due to the sophisticated modeling machinery, WGANTPP can predict the ranking fluctuations to some extent, thereby achieving significant performance boost in terms of SRCC, in most of the datasets. However, when the ranking fluctuations are quite high, for instance, in datasets with high rank diversity, like Nepal-Earthquake and Dem-Primary, WGANTPP fails to second CRPP, despite its strong predictive power. On the other hand, LMPP exploits the presence of such signals and performs best among the baselines. We also plot the variation in rank prediction performance (with respect to SRCC) with time in the test window for

Table 4 SRCC of the proposed and baseline algorithms on all datasets with 20% held-out test set

Datasets	SRCC						
	CRPP	LMPP	WGANTPP	RPP	Hawkes	SEISMIC	SpikeM
Halloween	**0.94** (6.82%)	0.87	*0.88*	0.75	0.51	0.12	0.77
Nepal-Earthquake	**0.97** (6.59%)	*0.91*	0.90	0.60	0.29	0.63	0.75
Dem-Primary	**0.93** (8.14%)	*0.86*	0.85	0.76	0.47	0.10	0.73
BBD	**0.99** (4.21%)	*0.95*	0.94	0.86	0.88	0.43	0.79
Amazon-SP	**0.96** (4.35%)	*0.92*	0.91	0.57	0.49	0.54	0.79
Amazon-LAP	**0.92** (4.54%)	*0.88*	0.84	0.76	0.54	0.71	0.81
Amazon-REF	**0.91** (4.60%)	*0.87*	0.81	0.72	0.52	0.43	0.78
Amazon-WM	**0.96** (3.22%)	0.92	*0.93*	0.74	0.81	0.43	0.75
Amazon-AC	**0.95** (2.15%)	*0.93*	0.91	0.82	0.87	0.45	0.78
Amazon-TV	**0.92** (5.75%)	0.84	*0.87*	0.75	0.82	0.52	0.74

Bold (italics) indicate the best (second best) predictor. Here CRPP reports the performance of the best among the variants of our proposal. Numbers in bracket in the CRPP cells indicate the percentage improvement from the second best predictor

Table 5 Average precision in jump detection for the proposed models and baselines

Datasets	Average precision						
	CRPP	LMPP	WGANTPP	RPP	Hawkes	SEISMIC	SpikeM
Halloween	**0.85** (14.86%)	*0.74*	0.65	0.54	0.36	0.32	0.35
Nepal-earthquake	**0.88** (44.26%)	*0.61*	0.54	0.34	0.23	0.40	0.44
Dem-primary	**0.91** (31.88%)	*0.69*	0.52	0.45	0.29	0.45	0.34
BBD	**0.92** (39.39%)	*0.66*	0.58	0.30	0.52	0.31	0.43
Amazon-SP	**0.88** (23.94%)	*0.71*	0.59	0.43	0.29	0.31	0.30
Amazon-LAP	**0.87** (24.28%)	*0.70*	0.64	0.35	0.26	0.41	0.43
Amazon-REF	**0.85** (16.44%)	*0.73*	0.62	0.41	0.31	0.42	0.37
Amazon-WM	**0.89** (23.61%)	*0.72*	0.67	0.32	0.44	0.33	0.41
Amazon-AC	**0.92** (21.05%)	*0.76*	0.63	0.37	0.53	0.38	0.31
Amazon-TV	**0.93** (25.68%)	*0.74*	0.65	0.43	0.59	0.31	0.45

Bold (italics) indicate the best (second best) predictor. Here CRPP reports the performance of the best among the variants of our proposal. Numbers in bracket in the CRPP cells indicate the percentage improvement from the second best predictor

our models as well as the baselines in Fig. 2. The observations here are the same as those for the MAPE$_N$ plot in the same figure.

Performance in Jump Detection Tables 5 and 6 report the average precision and recall of jump detection for LMPP, CRPP (best-performing variant), and baselines across all the datasets on the 20% held-out test set. We note that LMPP outperforms the baselines in terms of both precision and recall, followed by WGANTPP. This is because LMPP explicitly models the product-competitions, whereas WGANTPP does not. The stellar performance of CRPP in jump detection is expected since the underlying discriminator is especially geared towards capturing inter-product interactions. Since the loss functions used by all variants of CRPP deal with pairwise

Table 6 Average recall in jump detection for the proposed models and baselines

Datasets	Average recall						
	CRPP	LMPP	WGANTPP	RPP	Hawkes	SEISMIC	SpikeM
Halloween	**0.87** (16.00%)	*0.75*	0.66	0.32	0.35	0.34	0.37
Nepal-earthquake	**0.90** (28.57%)	*0.70*	0.63	0.33	0.30	0.52	0.57
Dem-primary	**0.93** (29.17%)	*0.72*	0.57	0.44	0.27	0.57	0.36
BBD	**0.94** (40.30%)	*0.67*	0.51	0.29	0.61	0.28	0.40
Amazon-SP	**0.86** (17.81%)	*0.73*	0.51	0.27	0.31	0.34	0.35
Amazon-LAP	**0.89** (25.35%)	*0.71*	0.63	0.33	0.30	0.32	0.37
Amazon-REF	**0.92** (22.67%)	*0.75*	0.67	0.41	0.29	0.47	0.35
Amazon-WM	**0.86** (19.44%)	*0.72*	0.61	0.31	0.52	0.31	0.36
Amazon-AC	**0.87** (14.47%)	*0.76*	0.58	0.43	0.54	0.36	0.41
Amazon-TV	**0.88** (18.92%)	*0.74*	0.62	0.41	0.28	0.32	0.46

Bold (italics) indicate the best (second best) predictor. Here CRPP reports the performance of the best among the variants of our proposal. Numbers in bracket in the CRPP cells indicate the percentage improvement from the second best predictor

Table 7 A comparison of $MAPE_t$ (%) between several variants of CRPP along with RMTPP as a baseline for $CRPP_{Gen}$ on all datasets with 20% held-out test set

Datasets	$MAPE_t$					
	$CRPP_{Unbiased}$	$CRPP_{Precision}$	$CRPP_{Recall}$	$CRPP_{AUC}$	$CRPP_{Gen}$	RMTPP
Halloween	5.47	**5.43**	5.51	*5.45*	5.93	6.52
Nepal-earthquake	6.53	**6.41**	6.57	*6.48*	7.30	7.67
Dem-primary	7.56	**7.41**	7.58	*7.47*	8.21	8.89
BBD	14.65	14.45	**14.34**	*14.40*	15.21	15.86
Amazon-SP	4.85	**4.70**	4.82	4.84	5.63	5.95
Amazon-LAP	*4.75*	4.83	**4.31**	4.91	5.45	5.77
Amazon-REF	4.86	*4.85*	**4.83**	5.19	5.93	6.46
Amazon-WM	4.84	**4.79**	4.82	*4.81*	5.95	6.78
Amazon-AC	4.83	4.67	**4.62**	*4.65*	5.89	6.52
Amazon-TV	4.63	**4.53**	4.88	*4.54*	5.63	6.33

Bold (italics) indicate the best (second best) predictor. Numbers in bracket indicate the percentage improvement from the second best predictor

interactions between products, they can more accurately predict when a product is going to be significantly promoted or demoted in rank relative to the other hashtags or products.

Performance Comparison Across the Variants of CRPP To have a better understanding about the workings of CRPP, we perform a comparative study among the performances of several variants of CRPP with different discriminative losses. We also consider the performance of $CRPP_{Gen}$, the generative-only version of CRPP in the context of RMTPP which serves a baseline for $CRPP_{Gen}$. This is illustrated in Tables 7, 8, 9, and 10. From Tables 7 and 8 we observe that precision and

Table 8 $MAPE_N$ (%) of several variants of CRPP along with RMTPP (used as a baseline for $CRPP_{Gen}$) on all datasets with 20% held-out test set

Datasets	$MAPE_N$ (%)					
	$CRPP_{Unbiased}$	$CRPP_{Precision}$	$CRPP_{Recall}$	$CRPP_{AUC}$	$CRPP_{Gen}$	RMTPP
Halloween2012	13.15	**12.145**	13.05	*12.84*	13.23	13.65
Nepal-earthquake	6.76	**6.70**	*6.82*	6.84	7.83	8.21
Dem-primary	7.86	*7.85*	**7.83**	7.89	8.93	9.45
BBD	14.83	14.67	**14.62**	*14.65*	16.89	17.42
Amazon-SP	4.76	**4.70**	*4.72*	4.81	4.85	4.93
Amazon-LAP	*4.34*	4.93	**4.21**	4.91	5.25	5.63
Amazon-REF	5.86	*5.25*	**4.85**	5.89	5.93	6.15
Amazon-WM	4.91	**4.79**	4.87	*4.81*	4.93	5.11
Amazon-AC	5.84	5.72	**4.82**	*5.65*	5.89	6.12
Amazon-TV	4.65	**4.53**	4.71	*4.57*	4.94	5.23

Bold (italics) indicate the best (second best) predictor

Table 9 SRCC of several variants of CRPP along with RMTPP (used as a baseline for $CRPP_{Gen}$) on all datasets with 20% held-out test set

Datasets	SRCC					
	$CRPP_{Unbiased}$	$CRPP_{Precision}$	$CRPP_{Recall}$	$CRPP_{AUC}$	$CRPP_{Gen}$	RMTPP
Halloween2012	0.87	0.92	*0.93*	**0.94**	0.74	0.71
Nepal-earthquake	*0.93*	**0.97**	0.83	0.85	0.75	0.71
Dem-primary	0.88	0.91	*0.92*	**0.93**	0.72	0.67
BBD	0.96	**0.99**	*0.98*	0.96	0.81	0.74
Amazon-SP	*0.92*	**0.96**	0.84	0.87	0.79	0.73
Amazon-LAP	0.87	0.89	**0.92**	*0.91*	0.78	0.72
Amazon-REF	0.84	0.87	*0.89*	**0.91**	0.73	0.66
Amazon-WM	0.89	0.91	*0.92*	**0.96**	0.84	0.78
Amazon-AC	0.81	**0.95**	*0.91*	0.86	0.82	0.75
Amazon-TV	0.87	*0.91*	0.89	**0.92**	0.85	0.74

Bold (italics) indicate the best (second best) predictor

recall turn out to be the best discriminative measures, when the performance is compared in terms of $MAPE_N$ as well as $MAPE_t$. However, from Table 9 we find that, in case of SRCC, AUC performs significantly better against all other ranking measures. This is because, AUC is a more powerful ranking measure than both precision and recall [12], and as a result it facilitates CRPP to capture the ranking dynamics more effectively than others. We also observe that $CRPP_{Unbiased}$ usually fares poorly due to its inability to capture the skewed ranking distribution present in most of the cases. $CRPP_{Gen}$ does not capture the product competition process at all, which severely affects its performance in terms of $MAPE_N$, $MAPE_t$, and SRCC. However, we can see that it consistently performs better than RMTPP in all cases, thereby justifying its incorporation as a component of CRPP in preference to using RMTPP [6] itself as a generator. From Table 10, we observe that $CRPP_{Precision}$

Table 10 Average precision and recall in jump detection for all variants of CRPP

Datasets	Average precision (average recall)			
	CRPP$_{Unbiased}$	CRPP$_{Precision}$	CRPP$_{Recall}$	CRPP$_{AUC}$
Halloween2012	0.76 (0.78)	**0.85** (0.81)	0.73 (**0.87**)	*0.82 (0.81)*
Nepal-earthquake	0.79 (0.78)	**0.88** (0.82)	0.81 (**0.90**)	*0.84 (0.86)*
Dem-primary	0.75 (0.74)	**0.91** (0.83)	0.82 (**0.93**)	*0.86 (0.85)*
BBD	0.73 (0.78)	**0.92** (0.82)	0.83 (**0.94**)	*0.87 (0.88)*
Amazon-SP	0.72 (0.68)	**0.88** (0.73)	0.74 (**0.86**)	*0.83 (0.82)*
Amazon-LAP	0.74 (0.76)	**0.87** (0.75)	0.71 (**0.89**)	*0.82 (0.81)*
Amazon-REF	0.73 (0.72)	**0.85** (0.74)	0.79 (**0.92**)	*0.82 (0.82)*
Amazon-WM	0.74 (0.73)	**0.89** (0.77)	0.75 (**0.86**)	*0.84 (0.82)*
Amazon-AC	0.76 (0.73)	**0.92** (0.76)	0.78 (**0.87**)	*0.85 (0.83)*
Amazon-TV	0.74 (0.72)	**0.93** (0.78)	0.76 (**0.88**)	*0.86 (0.85)*

Bold (italics) indicates best (second best) predictor

Table 11 Effect of competition: % improvement of MAPE$_N$ upon augmentation of learning-to-rank models

Datasets	% Improvement in MAPE$_N$					
	CRPP	WGANTPP	RPP	Hawkes	SEISMIC	SpikeM
Halloween	5.89	7.60	**12.89**	5.25	8.35	12.48
Nepal-Earthquake	**14.43**	7.65	7.88	2.58	11.13	3.29
Dem-Primary	12.32	**12.66**	8.28	3.78	6.45	11.23
BBD	**13.43**	7.21	9.79	6.40	5.35	8.22
Amazon-SP	3.09	**16.88**	9.69	5.65	5.38	9.81
Amazon-LAP	**19.81**	10.70	11.83	4.22	6.46	7.67
Amazon-REF	**18.21**	7.47	8.77	7.18	10.58	5.84
Amazon-WM	2.84	**9.08**	7.98	8.36	5.73	9.01
Amazon-AC	**18.17**	13.08	5.62	6.77	7.84	7.26
Amazon-TV	8.30	**18.99**	10.79	10.91	8.78	6.34

Bold indicates the best performer. The deep neural models provide the maximum boosts upon such augmentation, w.r.t. the other baselines

reports the highest precision in all datasets, while CRPP$_{Recall}$ reports the highest recall across all datasets. It is interesting to note that CRPP$_{AUC}$ always reports the second-highest precision and recall.

Effect of Competition Tables 11, 12, and 13 compare the effect of augmenting discriminative learning-to-rank templates across all the algorithms except LMPP and RMTPP. More specifically these tables report the improvement of the corresponding metrics for each generative model, on using the discriminative counter-parts. Table 11 shows that, upon the addition of such ranking losses, the deep neural models give a substantial accuracy boost in terms of MAPE$_N$, against the parametric counter-parts. This is because, upon addition of the discriminative modules, the deep models forecast the future events more accurately which is reflected in a

Table 12 Effect of competition: % improvement of SRCC upon augmentation of learning-to-rank models

Datasets	% Improvement in SRCC					
	CRPP	WGANTPP	RPP	Hawkes	SEISMIC	SpikeM
Halloween	27.03	6.02	11.94	18.60	**33.33**	18.46
Nepal-Earthquake	29.33	9.75	17.65	**26.09**	16.67	27.12
Dem-Primary	29.17	8.97	13.43	9.30	**42.86**	12.31
BBD	**22.22**	5.62	4.65	5.68	10.26	8.22
Amazon-SP	**21.52**	5.81	26.67	19.51	14.89	6.76
Amazon-LAP	17.95	6.33	13.43	**25.58**	10.94	6.58
Amazon-REF	**24.66**	6.58	7.46	20.93	10.26	16.42
Amazon-WM	14.28	4.49	4.22	8.00	13.16	**17.19**
Amazon-AC	15.85	7.06	10.81	14.47	**21.62**	16.42
Amazon-TV	8.23	11.54	15.38	12.33	**20.93**	19.35

Bold indicates the best performer. No algorithm consistently perfroms best across all the datasets

Table 13 Effect of competition: % improvement of average precision upon augmentation of learning-to-rank models

Datasets	% Improvement in average precision					
	CRPP	WGANTPP	RPP	Hawkes	SEISMIC	SpikeM
Halloween2012	30.77	14.03	20.00	13.89	14.28	**34.61**
Nepal-Earthquake	**50.77**	19.14	6.25	35.29	25.00	29.41
Dem-Primary	**40.00**	15.55	21.62	38.09	28.57	30.77
BBD	**41.54**	23.40	20.00	23.81	19.23	26.47
Amazon-SP	35.38	23.40	19.44	**38.10**	34.78	15.38'
Amazon-LAP	33.85	16.36	25.00	**36.84**	28.12	22.86
Amazon-REF	**30.77**	10.71	32.25	29.17	20.00	76.19
Amazon-WM	36.92	19.64	33.33	37.5	**43.48**	24.24
Amazon-AC	41.54	16.67	48.00	17.78	**65.22**	24.00
Amazon-TV	43.08	**44.44**	20.59	31.11	19.23	18.42

Bold indicates the best performer. No algorithm consistently performs best across all the datasets

smaller forecasting error. However, in terms of improvement of ranking losses, i.e., SRCC and average. precision, the best performer is not consistent across datasets (Tables 12 and 13). This is because, here we predict ranking fluctuations, by augmenting same discriminative module, as compared to Table 11 which measures the forecasting performance of models with different predictive principles.

Scalability We measured the time taken by the variants of CRPP to converge during training for (1) increasing number of competing products and (2) increasing number of training intervals (Fig. 3). We observe that the training times required by the variants of CRPP scale roughly sublinearly in both cases. Note that in practical situations, the number of competing products hardly goes beyond a couple of dozens. Also we observe that the number of training intervals is kept less than

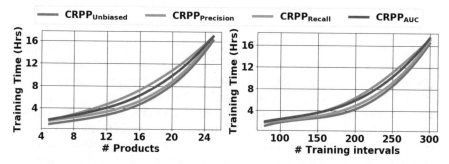

Fig. 3 Scalability plot for all CRPP variants showing time to converge with no. of products and no. of training intervals

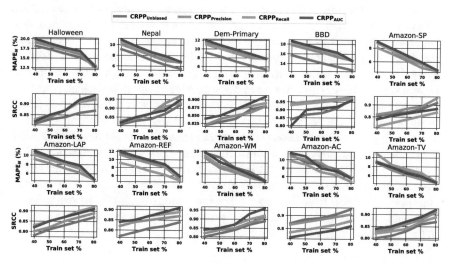

Fig. 4 Variation in performance of all variants of CRPP with respect to the $MAPE_N$ and SRCC metrics for different train-test splits across all datasets

300, to obtain significant rank fluctuations in individual intervals. Note that once the model is trained, which may be done offline, the time required to run it in a real-time production environment is insignificant.

Stability w.r.t. Training Set Size In addition to the above experiments we also studied the variation in performance of the variants of CRPP with respect to the $MAPE_N$ (forecasting) and SRCC (rank prediction) metrics with varying train-test splits (Fig. 4). The proportion of the training set in the available data was varied from 40% to 80% (our standard) in steps of 10% for all datasets. We observe that the performance of all variants of CRPP degrades roughly linearly with decreasing training set size, thus showing that CRPP is able to perform reasonably well in situations with limited training data.

4 Conclusion

In this paper, we propose LMPP, a novel point process driven framework, that unifies several realistic factors to model information dynamics. Such a unified approach does not only efficiently estimate hashtag popularity for which it is designed, but also gives an accurate prediction of the relative ranking of concurrent and competing items. Our second model, CRPP, is a novel probabilistic modeling framework for information visibility dynamics. Our approach unifies two major factors controlling information propagation—the competition between concurrent processes and their coupled nonlinear diffusion process. At the heart of our model lies an RNN guided generative framework, modeled in a discriminative adversarial setting. While the RNNs aim to model the complexity of the joint hashtag diffusion process, the discriminative system models the hashtag competition process. Such a model allows CRPP to accurately predict not only collective information flow but also the fluctuations in popularity ranking. Extensive experiments over several real datasets show that CRPP cements the baselines in predicting hashtag visibility dynamics. Our work opens many interesting avenues for research. For example, our current work does not explicitly capture any network structure in Twitter. Note that, in contrast to the traditional diffusion models—that only capture individual diffusion dynamics, our model aims to capture the coupled dynamics of information flow where each piece of information competes with others. The resulting social network is extremely sparse, disconnected and therefore our model as well as the competitors cannot extract any meaningful network signal from the data. However, if one is able to do so, it would help us to capture the visibility dynamics more accurately than our current model. Apart from this, it would be interesting to estimate the reputation of different brands using our model, which, we believe, is still a fertile area of research in the area of e-commerce, recommendation, etc.

References

1. Bao, P., Shen, H.-W., Jin, X., Cheng, X.-Q.: Modeling and predicting popularity dynamics of microblogs using self-excited Hawkes processes. In: Proceedings of the 24th International Conference on World Wide Web (2015)
2. Chen, W., Wang, Y., Yang, S.: Efficient influence maximization in social networks. In: Proceedings of the 15th ACM SIGKDD International Conference on Knowledge Discovery and Data Mining, pp. 199–208. ACM, New York (2009)
3. Chen, W., Yuan, Y., Zhang, L.: Scalable influence maximization in social networks under the linear threshold model. In: 2010 IEEE 10th International Conference on Data Mining (ICDM), pp. 88–97. IEEE, Piscataway (2010)
4. Crawford, M.B.: The World Beyond Your Head: On Becoming an Individual in an Age of Distraction. Farrar, Straus and Giroux, New York (2015)
5. De, A., Valera, I., Ganguly, N., Bhattacharya, S., Rodriguez, M.G.: Learning and forecasting opinion dynamics in social networks. In: Conference on Neural Information Processing Systems (2016)

6. Du, N., Dai, H., Trivedi, R., Upadhyay, U., Gomez-Rodriguez, M., Song, L.: Recurrent marked temporal point processes: embedding event history to vector. In: Proceedings of the 22nd ACM SIGKDD International Conference on Knowledge Discovery and Data Mining, pp. 1555–1564. ACM, New York (2016)
7. Farajtabar, M., Du, N., Gomez-Rodriguez, M., Valera, I., Song, L., Zha, H.: Shaping social activity by incentivizing users. In: Conference on Neural Information Processing Systems (2014)
8. Fujita, K., Medvedev, A., Koyama, S., Lambiotte, R., Shinomoto, S.: Identifying exogenous and endogenous activity in social media (2018). arXiv preprint arXiv:1808.00810
9. He, R., McAuley, J.: Ups and downs: modeling the visual evolution of fashion trends with one-class collaborative filtering. In: Proceedings of the 25th International Conference on World Wide Web, pp. 507–517. International World Wide Web Conferences Steering Committee, Geneva (2016)
10. Hua-Wei, S., Wang, D., Song, C., Barabasi, A.: Modeling and predicting popularity dynamics via reinforced Poisson processes. In: Proceedings of the Twenty-Eighth AAAI Conference on Artificial Intelligence (2014)
11. Joachims, T.: Optimizing search engines using clickthrough data. In: Proceedings of the Eighth ACM SIGKDD International Conference on Knowledge Discovery and Data Mining (2002)
12. Joachims, T.: A support vector method for multivariate performance measures. In: Proceedings of the 22nd International Conference on Machine Learning (2005)
13. Joachims, T.: Training linear SVMs in linear time. In: Proceedings of the 12th ACM SIGKDD International Conference on Knowledge Discovery and Data Mining, pp. 217–226. ACM, New York (2006)
14. Karimi, M., Tavakoli, E., Farajtabar, M., Song, L., Gomez-Rodriguez, M.: Smart broadcasting: do you want to be seen? In: KDD '16: Proceedings of the 22nd ACM SIGKDD International Conference on Knowledge Discovery in Data Mining (2016)
15. Kempe, D., Kleinberg, J.M., Tardos, E.: Maximizing the spread of influence through a social network. In: Proceedings of the Ninth ACM SIGKDD International Conference on Knowledge Discovery and Data Mining (2003)
16. Kempe, D., Kleinberg, J.M., Tardos, É.: Influential nodes in a diffusion model for social networks. In: Caires, L., Italiano, G.F., Monteiro, L., Palamidessi, C., Yung, M. (eds) Automata, Languages and Programming. ICALP 2005. Lecture Notes in Computer Science, vol. 3580. Springer, Berlin (2005)
17. Kobayashi, R., Lambiotte, R.: Tideh: time-dependent Hawkes process for predicting retweet dynamics. In: Proceedings of the Tenth International AAAI Conference on Web and Social Media (ICWSM 2016) (2016)
18. Matsubara, Y., Sakurai, Y., Prakash, B.A., Li, L., Faloutsos, C.: Rise and fall patterns of information diffusion: model and implications. In: Proceedings of the ACM SIGKDD International Conference on Knowledge Discovery and Data Mining (2015)
19. McAuley, J., Targett, C., Shi, Q., Van Den Hengel, A.: Image-based recommendations on styles and substitutes. In: Proceedings of the 38th International ACM SIGIR Conference on Research and Development in Information Retrieval, pp. 43–52. ACM, New York (2015)
20. McKelvey, K., Menczer, F.: Design and prototyping of a social media observatory. In: Proceedings of the 22nd International Conference on World Wide Web companion, WWW '13 Companion, pp. 1351–1358 (2013)
21. McKelvey, K., Menczer, F.: Truthy: enabling the study of online social networks. In: Proceedings 16th ACM SIGKDD Conference on Computer Supported Cooperative Work and Social Computing Companion (CSCW) (2013)
22. Medvedev, A.N., Delvenne, J.-C., Lambiotte, R.: Modelling structure and predicting dynamics of discussion threads in online boards. J. Complex Netw. **7**, 67–82 (2018)
23. Mei, H., Eisner, J.M.: The neural Hawkes process: a neurally self-modulating multivariate point process. In: Advances in Neural Information Processing Systems, pp. 6757–6767 (2017)

24. Mossel, E., Roch, S.: On the submodularity of influence in social networks. In: Proceedings of the Thirty-Ninth Annual ACM Symposium on Theory of Computing, pp. 128–134. ACM, New York (2007)
25. Ogata, Y.: On Lewis' simulation method for point processes. IEEE Trans. Inf. Theory **27**, 23–30 (1981)
26. Samanta, B., De, A., Chakraborty, A., Ganguly, N.: LMPP: a large margin point process combining reinforcement and competition for modeling hashtag popularity. In: Proceedings of the Twenty-Sixth International Joint Conference on Artificial Intelligence, pp. 2679–2685 (2017)
27. Samanta, B., De, A., Ganguly, N.: STRM: a sister tweet reinforcement process for modeling hashtag popularity. In: IEEE INFOCOM (2017)
28. Samanta, B., De, A., Ganguly, N.; STRM: a sister tweet reinforcement process for modeling hashtag popularity. In: INFOCOM 2017-IEEE Conference on Computer Communications, IEEE, pp. 1–9. IEEE, Piscataway (2017)
29. Spasojevic, N., Li, Z., Rao, A., Bhattacharyya, P.: When-to-post on social networks. In: Proceedings of the 21th ACM SIGKDD International Conference on Knowledge Discovery and Data Mining (2015)
30. Valera, I., Gomez-Rodriguez, M.: Modeling adoption and usage of competing products. In: Proceedings of the 2015 IEEE International Conference on Data Mining (ICDM) (2015)
31. Weston, J.: Support vector machine. Tutorial. http://www.cs.columbia.edu/~kathy/cs4701/documents/jason_svm_tutorial.pdf (2014). Accessed 10 May 2014
32. Xiao, S., Farajtabar, M., Ye, X., Yan, J., Yang, X., Song, L., Zha, H.: Wasserstein learning of deep generative point process models. In: Advances in Neural Information Processing Systems, pp. 3250–3259 (2017)
33. Zarezade, A., Upadhyay, U., Rabiee, H., Gomez-Rodriguez, M.: Redqueen: an online algorithm for smart broadcasting in social networks. In: WSDM (2017)
34. Zhao, Q., Erdogdu, M.A., He, H.Y., Rajaraman, A., Leskovec, J.: Seismic: a self-exciting point process model for predicting tweet popularity. In: Proceedings of the 21th ACM SIGKDD International Conference on Knowledge Discovery and Data Mining (2015)

Index

© Springer Nature Switzerland AG 2019
F. Ghanbarnejad et al. (eds.), *Dynamics On and Of Complex Networks III*,
Springer Proceedings in Complexity, https://doi.org/10.1007/978-3-030-14683-2

Printed in the United States
By Bookmasters